Praise for the book

"In his book, Mel Alexenberg navigates his artistic insight amid the labyrinthian complexities, explosions, and revolutions of the past forty years of art, tracing his way amid questions of science and religion, technology and environment, education, culture, and cosmos. Everyone will find his book full of new vantage points and vistas, fresh insights that give a uniquely personal history of artistic time that indeed points to new and open futures."

– Prof. Lowry Burgess, Dean and Distinguished Fellow of the Studio for Creative Inquiry, Carnegie Mellon University, Pittsburgh, PA

"This is a wonderful and important book. The author links the history of art to the important role played by various forms of thinking in the Jewish tradition and connects that to the emerging culture of digital expression. Brilliant insights and new ways of seeing make this a must-read for anyone interested in the intellectual history of images in the 21st century."

– Dr. Ron Burnett, author of *How Images Think* (MIT Press), President of Emily Carr University of Art and Design, Vancouver, Canada

"The author succeeds in opening a unique channel to the universe of present and future art in a highly original and inspiring way. His connection between ancient concepts (Judaism) and the present digital age will force us to thoroughly rethink our ideas about art, society and technology. This book is evidence that Golem is alive!"

– Prof. Michael Bielicky, Professor and Head of the Department of InfoArt/Digital Media, Hochschule fur Gestaltung /ZKM Center for Art and Media, Karlsruhe, Germany

"Mel Alexenberg, a very sophisticated artist and scholar of much experience in the complex playing field of art-science-technology, addresses the rarely asked question: How does the 'media magic' communicate content?"

– Prof. Otto Piene, Professor Emeritus and Director, MIT Center for Advanced Visual Studies, Massachusetts Institute of Technology, Cambridge, MA, USA

"Opens new vistas in the attempts to reconcile the newest developments in digital art and postmodern critical perspectives with the ancient concerns of the arts with the spiritual. It offers fresh perspectives in how we can learn from Greek and Jewish thought to understand the present era."

– Prof. Stephen Wilson, author of *Information Arts: Intersections of Art, Science, and Technology* (MIT Press) and Professor of Conceptual and Information Arts at San Francisco State University, San Francisco, CA, USA

"This book is simply a must-read analysis for anyone interested in where we and the visual arts are going in our future. Alexenberg has provided us with powerful new lenses to allow us to 'see' how postmodern art movements and classical Judaic traditions compliment and fructify one another as the visual arts are now enlarging and adding a spiritual dimension to our lives in the digital era."

– Dr. Moshe Dror, co-author of *Futurizing the Jews* (Praeger), President of World Network of Religious Futurists and Israel Coordinator of World Future Society

"This Hebraic-postmodern quest is for a dialogue midway on Jacob's ladder where man and God, artist and society, and artwork and viewer/participant engage in ongoing commentary."

– Dr. Randall Rhodes, Professor and Chairman, Department of Visual Art, Frostburg State University, Maryland, USA

"This book is amazing, so deep and insightful and full of sweet revelations at each turn of the page! It rocks the world and brings some desperately needed light."

– Dr. David Lazerson, author of *Sharing Turf* (Ballard), Inductee at the National Teachers Hall of Fame, USA

"Like the Torah itself that Alexenberg refers to regularly, the book is complex. He writes in a lively, engaging style [...] I found it informative, optimistic, and spiritually refreshing."

– Prof. Rob Harle, *Leonardo: Journal of the International Society for the Arts, Sciences and Technology*

"Alexenberg offers special insights into the postmodern nature of the Talmud's biblical consciousness as an open-ended living system. His argument is that the new paradigm of art must be of a structural and dynamic nature. Here, he quotes Allan Kaprow in urging a more 'lifelike art.' This has profound implications for art education."

– Dr. Jerome Hausman, *Arts and Activities*

"If Jacques Derrida had not preceded him, Alexenberg would be the Jewish Marshall McLuhan [...] Alexenberg's art and scholarship represents some of the most innovative work being made in both the Jewish and non-Jewish art worlds."

– Menachem Wecker, *Forward*

"Alexenberg's dynamic interplay of insider/outsider methodologies and exploration of the multiple relationships that exist between art, technology, and culture today is the highlight of this text. His combination of practice-based outcomes with scholarly negotiation of the topic presents a distinctive character to this research."

– Dr. Vince Dzeikan, *New Media & Society*

The Future of Art in a Postdigital Age

To my wonderful wife, Miriam, our children, grandchildren, and great-grandchildren.

From generation to generation, they will dwell in the Land of Israel
where the wilderness will rejoice over them,
the desert will be glad and blossom like a lily.
Her wilderness will be made like Eden
and her desert like a Divine garden.
Joy and gladness will be found there,
thanksgiving and the sound of music.

(Isaiah 35:1, 51:3)

The Future of Art in a Postdigital Age

From Hellenistic to Hebraic Consciousness

Mel Alexenberg

intellect Bristol, UK / Chicago, USA

First published in the UK in 2011 by
Intellect, The Mill, Parnall Road, Fishponds, Bristol, BS16 3JG, UK

First published in the USA in 2011 by
Intellect, The University of Chicago Press, 1427 E. 60th Street,
Chicago, IL 60637, USA

A catalogue record for this book is available from the British Library.

Cover designer: Holly Rose
Copy-editor: Emma Rhys
Typesetting: Mac Style, Beverley, E. Yorkshire

ISBN 978-1-84150-377-6

Contents

Author's Note for Postdigital Edition

When my publisher asked me to create an updated and enlarged edition of my 2006 book *The Future of Art in a Digital Age: From Hellenistic to Hebraic Consciousness*, it was clear to me that the book called for an updated title that reflected the emergence of a postdigital age. *Wikipedia*, itself an exemplary postdigital collaborative creation, explains that "postdigital" is a term used in the discourse of digital artistic practice at a time of our rapidly changing relationships with digital technologies and art forms more concerned with being human than with being digital. Reading this *Wikipedia* entry made me realize that much of the first edition of this book had already ventured beyond the digital to explore postdigital perspectives emerging from creative encounters between art, science, technology, and human consciousness. This Postdigital Edition continues that exploration.

I had first sensed a postdigital aura nearly a half-century ago when I programmed mammoth computers to make pictures that called out for the warm human touch of colorful pigments in molten beeswax sensuously flowing over a plotter's hard-edged digital drawings. This aura continued as I initiated interactive dialogues between human sensibilities and new technologies in the *LightsOROT: Spiritual Dimensions of the Electronic Age* exhibition that I created more than two decades ago in collaboration with Otto Piene at MIT's Center for Advanced Visual Studies for Yeshiva University Museum. As art editor of *The Visual Computer: International Journal of Computer Graphics*, I titled my 1988 digital art issue editorial "Art with Computers: The Human Spirit and the Electronic Revolution," an apt title today for a chapter in a book on postdigital art. My current works of participatory blogart and wikiart that mirror the living Talmud, the oldest on-going wikicreation that spans centuries and continents, continue my explorations of the human dimensions of new media.

When I checked *Wiktionary*, the wiki-based open content dictionary, for a definition of "postdigital" I found none. So I created one based upon my research for writing this Postdigital Edition of my book. I posted it on *Wiktionary* and added it to *Wikipedia*'s entry for "Postdigital." My act of collaborating in the creation of the world's most actively used dictionary and encyclopedia beautifully exemplifies the postdigital age. Now if you look for the *Wiktionary* definition, you will find mine.

> **Postdigital** (adjective), of or pertaining to art forms that address the humanization of digital technologies through interplay between digital, biological, cultural, and spiritual systems, between cyberspace and real space, between embodied media and mixed reality in social and physical communication, between high tech and high touch experiences, between visual, haptic, auditory, and kinesthetic media experiences, between virtual and augmented reality, between roots and globalization, between autoethnography and community narrative, and between web-enabled peer-produced wikiart and artworks created with alternative media through participation, interaction, and collaboration in which the role of the artist is redefined.

I added two chapters to this new edition: "Postdigital Perspectives: Rediscovering Ten Fingers" and "Wiki Perspectives: Multiform Unity and Global Tribes." They both elaborate on my *Wiktionary* definition through theoretical discussions and descriptions of exemplary artworks.

The black and white photographs in the first edition have been dropped in favor of inviting my readers to see photographs of artworks in full color posted on my website (www.melalexenberg.com), on my *Artiststory*, *Postdigital Art* and *Wikiartists* blogs, and on my *Postdigital Art* Facebook group wall. This digital solution not only offers access to many more photographs than were printed in the first edition, but allows for continual updating with new images. So take breaks from reading this book and enter into cyberspace to enrich your encounters with my words through viewing the colorful images posted on my website and blogs.

Mel Alexenberg, School of the Arts, Emuna College, Jerusalem, Israel

Introduction

Postmodern Paradigm Shift: *From Hellenistic to Hebraic Consciousness*

If Israelite thinking is to be characterized, it is obvious first to call it dynamic, vigorous, passionate, and sometimes quite explosive in kind; correspondingly Greek thinking is static, peaceful, moderate, and harmonious in kind.[1]

Thorleif Boman

The Greeks and the Jews are the two peoples whose worldviews have most influenced the way we think and act. Each of them from angles so different has left us with the inheritance of its genius and wisdom. No two cities have counted more with Mankind than Athens and Jerusalem. Their messages in religion, philosophy, and art have been the main guiding light in modern faith and culture.[2]

Winston Churchill

The worldview of ancient Greece revived in Renaissance Europe dominated Western art and architecture until the rise of modernism. The transition to modernism and postmodernism in American art and architecture represents a paradigm shift from the Hellenistic to the Hebraic roots of Western culture exemplified by the two Guggenheim art museums – Frank Lloyd Wright's museum in New York and Frank Gehry's museum in Bilbao, Spain. An analysis of these two major works of American architecture provides an introduction to the contemporary confluence between Hebraic consciousness and postmodern art in a postdigital age.

In his seminal work, *Hebrew Thought Compared with Greek,* Norwegian theologian Thorleif Boman analyzes Hellenistic and Hebraic consciousness and compares them. He emphasizes the dynamic, vigorous, passionate, and action-centered characteristics of Hebraic consciousness in contrast to the static, peaceful, moderate, and passive Greek consciousness. Boman notes that biblical passages concerned with the built environment always describe plans for construction without any description of the appearance of the finished structure. Noah's ark is presented as a detailed building plan. How the ark looked when it set sail is never described. The Bible has exquisitely detailed construction instructions for the

Tabernacle without any word picture of the appearance of the completed structure. Indeed, the Tabernacle was made of modular parts and woven curtains, came apart like Lego, was set on a wagon, moved through the desert from site to site, deconstructed and reconstructed each time. Its modest tent-like design, human scale, and active life was quite different from the immovable monumental marble temples on the Acropolis.

A biblical structure of consciousness in architecture emphasizes temporal processes in which space is actively engaged by human community rather than presenting a harmoniously stable form in space. Architectural theorist, Bruno Zevi, compares the Hebraic and Greek attitudes toward architecture in his essay, "Hebraism and the concept of space-time in art."

> For the Greeks a building means a house-object or a temple-object. For the Jews it is the object-as-used, a living place or a gathering place. As a result, architecture taking its inspiration from Hellenic thought is based on colonnades, proportions, refined moulding, a composite vision according to which nothing may be added or eliminated, a structure defined once and for all. An architecture taking its inspiration from Hebrew thought is the diametric opposite. It is an organic architecture, fully alive, adapted to the needs of those who dwell within, capable of growth and development, free of formalistic taboo, free of symmetry, alignments, fixed relationships between filled and empty areas, free from the dogmas of perspective, in short, an architecture whose only rule, whose only order is change.[3]

Frank Lloyd Wright's Guggenheim Museum

In *Frank Lloyd Wright: A Study in Architectural Content,*[4] art historian Norris Kelly Smith explained Wright's originality and genius in terms of Boman's comparison between Hebrew and Greek patterns of thought. Since Wright was well versed in the Bible as the son of a Unitarian minister, he internalized the biblical message of freeing humanity from enslavement in closed spaces and expressed this freedom in his architectural design. Smith emphasized that Wright imbued the field of architecture, conditioned by two thousand years of Greco-Roman thought, with Hebrew thought. Wright detested Greek architecture both in its content and in its forms. He was critical of the neo-classical rhetoric employed by American architects who studied at the Ecole des Beaux-Arts in Paris. Wright sought to create a new architecture to echo the biblical call inscribed on the Liberty Bell in Philadelphia: "Proclaim liberty throughout all the land unto all the inhabitants thereof" (*Leviticus 25:10*). He wanted American architecture to assert its cultural independence from Europe.

The connection between the exodus of the Israelites from Egyptian slavery and the American experience as a rebellion against European tyranny was clear to America's founding fathers. On July 4, 1776, the Continental Congress formed a high-powered committee, made up of Thomas Jefferson, Benjamin Franklin and John Adams, to propose a seal and motto for the newly independent United States of America. They proposed a seal depicting the

Israelites escaping to freedom from bondage under Pharaoh through the divided waters of the Red Sea, with Moses standing on the shore extending his hand over the sea, causing it to overwhelm the Egyptians. The proposed motto: "Rebellion to Tyrants is Obedience to God."[5] Fourteen years later, George Washington wrote a letter to the Hebrew Congregation of Savannah repeating the same biblical message of freedom:

> May the same wonder-working Deity, who long since delivered the Hebrews from their Egyptian oppressors, planted them in the promised land, whose providential agency has lately been conspicuous in establishing these United States as an independent nation, still continue to water them with the dews of heaven and make the inhabitants of every denomination participate in the temporal and spiritual blessings of the people whose God is Jehovah.[6]

It is significant that the nation founded on the principles of "life, liberty, and the pursuit of happiness" became the center of the shift from the Hellenistic to the Hebraic worldview in the arts. Dynamic forms of art and architecture symbolizing life and liberty blossomed on American soil. Frank Lloyd Wright exemplified this blossoming. His spiral museum invites a living response. When I had asked my children what they remembered most from their visits to the Guggenheim, they enthusiastically reminisced about running down the ramp and being high up looking over the fence into the center atrium. It is not a box for rectangular pictures set in static space, it is a lively place to be engaged over time. The exhibitions I saw there that worked best were shows about movement: Alexander Calder's mobiles were moving around the spiral to create a circus of color. Yaacov Agam's kinetic and dialogic art changed with the movement of the viewers in his *Beyond the Visible* show, and Jenny Holzer's ruby light word messages on a running electronic signboard flashed their way up the spiral ramp. The motorcycle show was right on the mark.

The spiral is one of the major life forms in nature: from DNA, to a nautilus shell, to the growth pattern of palm fronds. It is also one of the major symbols of the Hebraic mind. Jews are called *am haSePheR*, usually translated "People of the Book." But *SePheR* is a word written in the Torah scroll itself long before the invention of codex type books. *SePheR* means spiral scroll. It is spelled *SPR*, the root of the word "SPiRal" in numerous languages, ancient and modern. Jews, then, are People of the Spiral. In *kabbalah*, down-to-earth biblical mysticism, the *SePhiRot* are emanations of divine light spiraling down into our everyday life. And the English words "SPiRitual" and "inSPiRation" share the SRP root from the Latin *SPiRare*, to breathe.

In Judaism, form gives shape to content. The medium is an essential part of the message. Rather than the modernist viewpoint of art as "the language of forms," Judaism shares postmodernism's emphasis on "the ideas their forms might disclose."[7] Weekly portions of the first five books of the Bible in the form of a Torah scroll are read in synagogue. The symbolic significance of the spiral form is so strong that if a Torah scroll is not available in synagogue, the Bible is not publicly read at all. The exact same words printed in codex book form convey the wrong message. If the divine message encoded in the Torah is trapped between two rectilinear covers, it loses its life-giving flow. The message of the Torah must not be enslaved in the rectangle. It must have the infinite flow of a Mobius strip where the final letter of the Torah, the lamed of yisrae***L*** (Israel) connects to the first letter, the bet of ***B***'reshit (in the beginning). Lamed bet spells the word for "heart." The heart of the Torah is where the end connects to the beginning in an endless flow. Form and content join together to symbolize the essence of Jewish values. The Bible encoded in a flowing scroll form provides a clue as to the nature of biblical consciousness as an open-ended, living system.

> Wright's helicoidal shaping of the Guggenheim Museum's cavity in New York represents the victory of time over space, that is, the architectural incarnation of Hebrew thought, even more significant because it was fully realized by a non-Jew. Like Schonberg's music, Wright's architecture is based on linguistic polarity, emancipated dissonance, contradiction; it is once Expressionistic and rigorous; it applies Einstein's concept of "field"; it is multidimensional; it extols space by demolishing all fetishes and taboos concerning it, by rendering it fluid, articulated so as to suit man's ways, weaving a continuum between building and landscape. In linguistic terms, this means a total restructuring of form, denial of any philosophical a priori, any repressive monumentality: action-architecture, aimed at conquering ever more vast areas of freedom for human behavior.[8]

Frank Gehry's Guggenheim Museum

In creating the Bilbao Guggenhiem, Frank Gehry moved beyond Wright to a more powerful realization of the Hebraic mindset that Boman describes as dynamic, vigorous, passionate, and sometimes quite explosive in kind. It started in Canada when young Frank Goldberg (his father changed the family name when they moved to LA) would play with the live carp swimming in his grandmother's bathtub. Gehry often told the story that every Thursday his grandmother would buy fish and keep them in the bathtub until Friday when she prepared gefilte fish for the Sabbath meal. The vigorous body motions of swimming fish seen from above gave Gehry his vocabulary for the dynamic shape of his museum. Fish are one with their environment. They must stay in constant motion in it to stay alive. Oxygen carrying water must be kept moving over their gills for them to breathe. To stop motion is to die.

Gehry's method of working is creative play with dynamic forms. He starts with spontaneous scribble sketches that become forms that he moves and reshapes in a dynamic interplay between computer-generated 3D CAD graphic models and physical models in real space.

> Over the years, Gehry has cultivated a highly personal studio practice of working with models, because it permits impossibly cantilevered parts and vertiginous piles of volumes in fluid transformation. As he began to shape buildings from mobile parts, his sense of space transcended Cartesian notions. This special sense defies verbal definition, but it

> might be compared with the sensation of moving bodies in a medium akin to water. To the extent that his buildings arrest volumes in continuous motion (and transformation), time becomes their formative dimension.[9]

As an integral part of education for an architecture of time and motion, Gehry takes his students on ice in full hockey gear to interact with each other and their environment in rapid movement. Like fish in water, skaters standing still on ice are unstable. Swift motion creates balance. The same concept of stability in motion is sensed in seeing the "fish-scale" titanium skin on the Bilbao museum that makes it look like a futuristic airplane. Airplanes must move through their air medium in order to fly. Stopping motion in midair leads to crashing and death. He sets the bodies of his buildings in motion as a choreographer does with dancers. "One need only observe Gehry's manner of drawing to gain an immediate impression of his way of thinking: the pen does not so much glide across the page as it dances effortlessly though a continuum of space."[10] His studio practice appears like a performance rehearsal. His knowledge of performance art, his collaborations with artists, and his planning with artists led to spaces at the Bilbao Guggenheim uniquely suited for the presentation of alternative forms of art.

Gehry creates a dynamic flow between the building and its waterfront site and between the visitor and continually unfolding spaces. While jutting out over the water, the huge flowing fish-like building uses a combination of water-filled pools and the river to create an energetic interplay between building and site. Its full aerodynamic form can be seen from the other side of the river. Crossing the bridge and approaching the building transforms the experience of this monumental sculptural form into a more intimate encounter. Shifting viewpoints confuse the building and its environment as well as interior and exterior spaces. Movement through and around Gehry's museum always provides fresh encounters and new ways of seeing.

Hebraic consciousness and postmodernism

Roots of the confluence between Hebraic consciousness and postmodern art can be found in the account of the building of the Tabernacle in the Sinai desert under the direction of the artists Betzalel and Oholiav. "See, I have called by name: *Betzalel ben Uri ben Hur*, of the tribe of Judah. I have filled him with a divine spirit, with wisdom, understanding, and knowledge, and with the talent for all types of craftsmanship" (*Exodus 31:2*). The literal translation of this artist's name is: "In the Divine Shadow son of Fiery Light son of Freedom." It honors the artist's passion and freedom of expression paralleling the spirit of modernism. The Bible also describes Betzalel's partner, Oholiav. "I have assigned with him *Oholiav ben Aẖisamakh* of the tribe of Dan, and I have placed wisdom in the heart of every naturally talented person" (*Exodus 31:6*). Oholiav's full name means "My Tent of Reliance on Father, Son, and My Brother," integrating the contemporary with its past and future. Father, son, and brother stand together with the artist in a common tent in mutual support of one another. Betzalel represents the psychological power of the artist and Oholiav the sociological impact on community. Together they symbolize the postmodern value of harnessing the passion and freedom of the artist to nurture intergenerational collaboration in building a shared environment of spiritual power.

Hebraic consciousness shares with postmodernism a dynamic, creative, playful consciousness that promotes the interplay between multiple perspectives and alternating viewpoints from inside and from outside. Contemporary Hebraic consciousness is shaped by creative encounters with Torah, the wellsprings of Judaism, over millennia. *Torah* is a word that has multiple meanings in Jewish tradition. Torah can be used specifically for the Five Books of Moses, the first five books of the Bible as written in the Torah scroll. It also can be used generically to refer to the entire body of Jewish learning. It can include the whole *TaNaKh*, an acronym for the full Hebrew Bible: the Pentateuch (***T****orah*), the Prophets (***N****eviim*), and the Holy Writings (***K****etuvim*: Psalms, Proverbs, Song of Songs, etc.). It includes both the Written Torah and the Oral Torah. The *Tanakh* is the Written Torah. The Oral Torah contains the manifold perspectives handed down from generation to generation since it was received at Mount Sinai. The intergenerational dialogue of rabbis creatively exploring Torah concepts in relation to everyday life is recorded in the Talmud. In counterpoint to the exoteric Talmud is the esoteric *kabbalah* that reveals the deepest levels of biblical consciousness by exploring parallel creative processes – divine and human. It is significant that the word *Torah* itself has multiple meanings congruent with the multiple perspectives of Hebraic consciousness.

The Talmud teaches us to see seventy facets of every biblical passage since the Hebrew word for "eye," *ayin*, is also the name of the Hebrew letter with a numerical equivalent of seventy. Jews traditionally study in dialogue with a learning partner continually seeking new and alternative ways of understanding the text, moving past the surface, beyond the literal meaning of the words, drawing close to the sacred text through creative play. The millennia-old symbol of Judaism is the *menorah*, a tree-like candelabrum opening up into multiple branches. The Torah is called a "Tree of Life." When it is placed in the ark in the synagogue, the congregation sings: "Its ways are ways of pleasantness and all its paths are peace" (*Proverbs 3:17*). No one way and no single path, but rather many ways and multiple paths.

This negation of a single-point perspective is evident in both postmodern art and in the emergence of new scientific paradigms as exemplified by Ilya Prigogione's conceptual model of the thermodynamics of nonequilibrium systems for which he was awarded the Nobel Prize. He explains in *Order Out of Chaos: Man's New Dialogue with Nature*[11] that the traditional science of the age of the machine tended to emphasize stability, order, uniformity, equilibrium, and closed systems. The transition from an industrial society to a high-technology society in which information and innovation are critical resources, brought forth new scientific world models that characterize today's accelerated social change: disorder, instability, diversity, disequilibrium, nonlinear relationships, open systems, and a heightened sensitivity to the flow of time.

This paradigm shift in science is echoed in the arts by Peter Weibel in *net_condition: Art and Global Media*: "Modern art created the aesthetic object as a closed system as a reaction to the machine-based industrial revolution. Post-modernism created a form of art of open fields of signs and action as a reaction to the post-industrial revolution of the information society."[12]

Down-to-Earth Spirituality

Complimenting modernism's movement of art to a higher spiritual realm of pure form and color, postmodernism is moving art down into everyday life and out across the planet. Kandinsky explored the spiritual nature of the emerging modern art movements at the beginning of the 20th century in his classic book, *Concerning the Spiritual in Art.*[13] He saw modern art as movement away from representation of the material world to a more spiritually elevated world of abstraction. He symbolized this spiritual ascent by a moving triangle with its apex leading it *forwards and upwards*. Postmodern art, on the other hand, strives for movement *downward and outward*, symbolized by a second triangle moving into the future through the wisdom of the past with the apex pointing downwards. These two triangles intertwined form the Star of David that symbolizes the dynamic integration of both up and down movements, like the biblical image of angels ascending and descending on Jacob's ladder linking heaven and earth. It symbolizes the shift of art from the Hellenistic to the Hebraic roots of Western culture.

In contrast to Greek thought honoring static, uniform, space-centered, closed systems in which the spiritual exists above the mundane, Hebraic thought celebrates dynamic, multiform, time-centered, open systems in which spirituality is drawn down into every part of our daily lives. Rather than a quest for purity of form in some heavenly realm, Judaism seeks to reveal spirituality in the rough complexities of earth-bound living. The centrality of this down-to-earth spirituality in Jewish consciousness is expressed by two of the towering figures in 20th century Jewish thought: H̲asidic leader Menachem M. Schneerson, the Lubavicher Rebbe; and Joseph B. Soloveitchik, professor at the rabbinical school at Yeshiva University. Rabbi Schneerson teaches:

> It is not enough for the Jew to rest content with his own spiritual *ascent*, the elevation of his soul in closeness to G-d, he must also strive to *draw spirituality down* into the world and into every part of it – the world of his work and his social life – until not only do they not distract him from his pursuit of G-d, but they become a full part of it.[14]

The Lubavitcher Rebbe takes a mystical path through the esoteric world of *kabbalah*. His H̲asidic route aims at releasing the holy sparks hidden in our mundane world. On the other hand, Rabbi Soloveitchik takes a highly intellectual path through the exoteric world of Talmudic learning. His is the path of the Mitnaged that develops a rational legalistic philosophy critical of mystical religion. He emphasizes *halakhah*, the practical rules of Jewish conduct. That both rabbinic leaders arrive at the same place through routes poles apart parallels the simultaneity of multiple paths that characterizes postmodernism. Rabbi Soloveitchik teaches:

> Judaism does not direct its glance upward but downward. The Halakhah does not aspire to a heavenly transcendence, nor does it seek to soar upon the wings of some abstract, mysterious spirituality. It fixes its gaze upon concrete, empirical reality [...] The Halakhah is not hermetically enclosed within the confines of cult sanctuaries but permeates into every nook and cranny of life. The marketplace, the street, the factory, the house, the meeting place, the banquet hall, all constitute the backdrop of religious life.[15]

In his acclaimed novel, *City of God*, E. L. Doctorow summarizes the confluence between postmodern experience and Jewish consciousness in a postdigital age in an elegant literary formulation. The main character in his novel, an Episcopal priest in the process of converting to Judaism, explains to his friend:

> If there is a religious agency in our lives, it has to appear in the manner of our times. Not from on high, but a revelation that hides itself in our culture, it will be ground-level, on the street, it'll be coming down the avenue in the traffic, hard to tell apart from anything else. It will be cryptic, discerned over time, piecemeal, to be communally understood at the end like a law of science.

His friend's response: "Yeah, they'll put it on a silicon chip."[16]

Multiple perspectives

The single-point perspective in Renaissance paintings gives the illusion of a three-dimensional space receding into the picture. In Giotto's painting *Madonna Enthroned*, all lines converge at one point, at the baby sitting on his mother's lap. The Madonna is placed center stage with admiring people on both sides. No people are placed in front of the stage leaving a space for the viewer to stand outside the painting looking in. The same outsider perspective was also the guiding principle of early empirical science. "It is the illusion of the visualist," writes the historian of science A. D. Richie, "that he can observe completely objectively, like a spectator who is not performer in a game."[17]

Modernism overthrew the single-point perspective of the Renaissance renewal of Hellenism in favor of seeing from multiple perspectives. Picasso and Braque fragmented the picture plane in their Cubist paintings to simultaneously present diverse perspectives to the viewer. At the same time that modernism was developing in art, scientists were proposing the multiple perspective of seeing light as both particle and wave. The Uncertainty Principle was proposed to explain that it is impossible to determine simultaneously the position and velocity of an electron. Although modernist art and the science of the same era replaced uniperspective illusions with dynamic multi-perspective images, they maintained the viewpoint of an outsider.

Postmodernism expanded the multi-perspective experience by coupling the viewpoint of the outsider with the viewpoint of the insider looking both inward and outward. The passive observation of the outsider is extended to the active participation of the insider in postmodern art. In both natural and social sciences, it is being recognized that the observer can change what is being studied by the very act of observation. The dialogic relationship between inside and out in Jewish consciousness is revealed in the Hebrew words *p'nim* and *panim*, which mean inside and outer face. Although vocalized differently, they are spelled with the same Hebrew letters *PNIM.*[18]

This book follows postmodernism's lead by exploring Jewish consciousness and art in the postdigital age from the perspectives of both outsider and insider. It is shaped by the interplay between my being a creative artist and observant Jew with an insider's perspective as well as being an academic scholar teaching art and Jewish thought with the objectivity of an outsider. My writing reflects my being a third-generation American who lives in Israel speaking English to my grandchildren living in New Hampshire and speaking Hebrew to my grandchildren living in Israel's Negev desert.

The multicultural dynamics of postmodern society not only exists between countries but within them. Photographs of me as a bearded Jew in front of three buildings in Brooklyn that have had great significance in my life became a conceptual artwork that explores cultural change in America. I was born in the Brooklyn Jewish Hospital, now Interfaith Hospital. I celebrated my becoming a bar mitzvah at my Uncle Morris' storefront synagogue on Coney Island Avenue, now a Pakistani mosque. And I married the great-granddaughter of the Chief Rabbi of the Netherlands in a Jewish wedding hall on Eastern Parkway, now an African-American Baptist church. In Israel, I taught my classes at Ariel University in Hebrew, the living biblical language that shapes multi-perspective Jewish consciousness. My students included Amharic speaking Jews from Ethiopia, Russian speaking Jews from different areas of the former Soviet Union, Arabic speaking Arab students, the children of Arabic speaking Jews from all the lands of North Africa and the Middle East, Spanish speaking Peruvian Indians who converted to Judaism and live in settlements in Samaria, Spanish speaking Jews from South America, and Hebrew speaking Jews whose families have lived in the Land of Israel for generations.

The chapters "Semiotic Perspectives" and "Morphological Perspectives" emphasize the outsider's viewpoint by employing semiotic and morphological analysis of art forms and the cultures that shape them in relation to Jewish consciousness. Semiotics, the theory of signs and how they create significance, provides a conceptual framework for redefining art. It creates categories of representational and presentational art forms, from Hellenistic iconic representation to Hebraic dialogic presentation. Morphological perspectives are explored using morphological hermeneutics as a method for studying civilizations as structures of consciousness. The comparison of space-time morphologies in mythological, logical, and ecological cultures traces how postmodernism developed art forms that mirror the structure of Jewish consciousness. Tsion Avital develops this structural approach to understanding the evolution of consciousness in art in his book *Art Versus Noart: Art Out of Mind.* He proposes that the century of modern art represents an attempt to deconstruct millennia of figurative art to free up the field of art so that a new reconstructive era of art can emerge. "The new paradigm of art must be of a structural and dynamic character, in contrast to the figurative paradigm, which was content-oriented and static."[19]

In *Consciousness and Culture*,[20] Eric Mark Kramer discusses how methodologies for studying cultures are not neutral. The methodologies themselves are cultural constructs that color the way we see other cultures. We need to recognize the limitations of the outsider's viewpoint that can attempt to fit cultural phenomena into alien categories. As an outsider, Boman's academic study of biblical consciousness in *Hebrew Thought Compared with Greek* employs methodologies and a viewpoint derived from Hellenistic thought. In addition to using Hellenistic outsider methodologies, I am interested in using Hebraic insider methodologies to study Hebraic biblical consciousness. Dynamic interplay between these two alternative methods of exploring art, technology, and culture shapes my thoughts throughout the book. My credentials as an outsider includes teaching the doctoral research methods course at Columbia University and the graduate seminar "Art, Technology, and Culture" at MIT's Center for Advanced Visual Studies. My outsider's access to semiotic and hermeneutical methods of analysis is counterbalanced by my insider's perspectives as a creative artist and observant Jew.

The chapters "Kabbalistic Perspective" and "Halakhic Perspective" present an insider's viewpoint based upon the personal experience of an active participant in Jewish life and in artistic creation. My insider's access to these methodologies comes from my being a descendent of the Israelites who stood at the foot of Mount Sinai to receive the Torah, both in its written form and the oral form passed down to me from generation to generation. The Oral Torah is wisdom emerging from the white spaces between the Hebrew letters of the Bible as the exoteric *halakhah* and the esoteric *kabbalah.* The halakhic perspective teaches how to live our everyday life in Jewish consciousness, how to draw spirituality down into our gross material world, and how to beautify our lives by relating to others with loving-kindness. The kabbalistic perspective provides a symbolic language and conceptual schema for exploring two parallel creative processes – human and divine. Both the description of the artistic personality and *kabbalah* are derived from the same biblical passage (*Exodus 31:2*). "Wisdom, understanding, and knowledge" are both the artist's cognitive traits and the first stages in the kabbalistic model of the creation of the universe.

The changes in this updated and enlarged edition are reflected in the change of title from *The Future of Art in a Digital Age* to *The Future of Art in a Postdigital Age.* Although the interweaving of outside and insider viewpoints throughout this book reveals a postdigital perspective, I have added two new chapters to emphasize the shift to a postdigital age that addresses the humanization of the digital technologies. "Postdigital Perspectives:

Rediscovering Ten Fingers" explores the emergence of innovative art forms that explore movement between cyberspace and real space; interplay between visual, haptic, and kinesthetic experiences; interactions between roots and globalization; interrelationships between digital, biological, and cultural systems; and dialogue between electronic, material, and spiritual realms. "Wikiperspectives: Local and Global Collaborations" surveys artworks created through participation, teamwork, interaction, and collaboration in which the role of the artist is redefined. Exemplary artworks realized through group efforts in local public spaces are discussed in relation to web-enabled peer-produced artworks of a global scale.

Imagery concepts

There is significant relevance of a central facet of Hebraic methodology to new directions in postdigital art. In contrast to Greek thought that manipulates abstract concepts, Hebraic thought uses imagery concepts drawn from everyday life experiences, concepts that are concrete yet metaphorical. Imagery concepts provide raw materials for artists to make art in a postmodern era. In "The Imagery Concept in Jewish Thought," Rabbi Adin Steinsaltz writes:

> Modern Western thought is characterized by an extensive use of abstract concepts that exist and operate within a more general abstract system. Jewish thought on the other hand, has, with very few exceptions, done without them. Abstract concepts are not to be found in the Bible [...] It is, of course, quite impossible to manipulate complex and abstract ideas without concepts, and Jewish sources do employ them. The concepts, used, however, are not abstract but rather plastic and depictive, and they communicate their meaning like a metaphor or visual representation. In other words, Jewish thought uses pictorial or imagery concepts instead of abstract concepts [...] Perhaps the best definition is that the imagery concept is a special case, a concrete and characteristic example, of a certain category of abstraction.[21]

"The blow of a hammer," for example, is an imagery concept used in Jewish thought to encompass the entire range of activities that would be expressed in abstract concepts like "the completion of a task" or "putting on finishing touches." "The blow of the hammer" is more than a symbolic phrase or metaphor that compresses a wide range of phenomena into a single term. It includes the special case as a tangible event within the general category.

David Gelernter, Yale University computer science professor and observant Jew has written ground-breaking books on both the aesthetics of computer science (*The Muse and the Machine: Computerizing the Poetry of Human Thought*[22]) and theme-images in Jewish life (*Judaism: A Way of Being*[23]). In his paper, "Judaism Beyond Words," Gelernter echoes the words of Rabbi Steinsaltz that Jewish thought uses imagery concepts rather than abstract concepts:

> Judaism's themes are conveyed in words. But often, the point of the words is to convey a picture. Christianity's central theme is a picture – a picture that has been repeated millions of times, in millions of ways. But Judaism's pictures are different. They are too allusive or abstract to paint, and yet they are pictures nonetheless. To understand the internal life of Judaism, you must be able to read the pictures. Some people have the wrong impression that Judaism is an "unvisual" or even anti-visual religion. It does strongly forbid the worship of images – yet Judaism may be the most "visual" of any Western religion or intellectual system.[24]

Gelernter proposes four overriding theme-images that permeate all of Jewish thought and experience: *Separation* in time and space as a process to counter entropy. *Veil* as the sacred screen that hides God from man and man from God while drawing them together. *Perfect Asymmetry* between man and woman in which two differently formed parts are put together to make a perfect whole. *Inward Pilgrimage* in the relentless spiraling inward to reveal God's still small voice within each person. He creates a poetic vision of the vibrant interaction between these theme-images:

> We are left with four images as we circle the city of Judaism. From the north, the Torah scroll held high and wide open, to the sound of rushing water; from the east, a large tallit by itself, and the sound of a shofar; from the south, a man and woman embraced ("Let him kiss me with his mouth's kisses! – your love is better than wine," Song of Songs 1:2) as someone blows out the lamp; from the west, Abraham's face at the moment he catches sight of Mount Moriah, to the sound of a still, small voice – rustling satin, or the long sigh (faint cymbals) of ice-covered branches trembling in a breeze. Then when we close our eyes and look again, and all four images have been consumed by a burning thornbush and the snap of flames.[25]

Talmud and the Internet

The major media systems in Jewish culture and the postdigital era share a common structure. Both the typographical design of the Talmud, the major work of Jewish law and lore, and hypertext linking in the design of the Internet are structured so that they facilitate and encourage creative, associative, and multiple perspectives. The confluence between the media ecologies of the Talmud and the Internet generate common Jewish and postdigital age structures of consciousness. In contrast, the single-point perspective and unilinear structure of Hellenistic consciousness revealed through Renaissance art and the industrial age Gutenberg Bible produce an obsolete structure of consciousness alien in an era of globalization and digital technologies. This book explores how fresh directions for art in the future can be derived from the shared structure of Jewish and digital media.

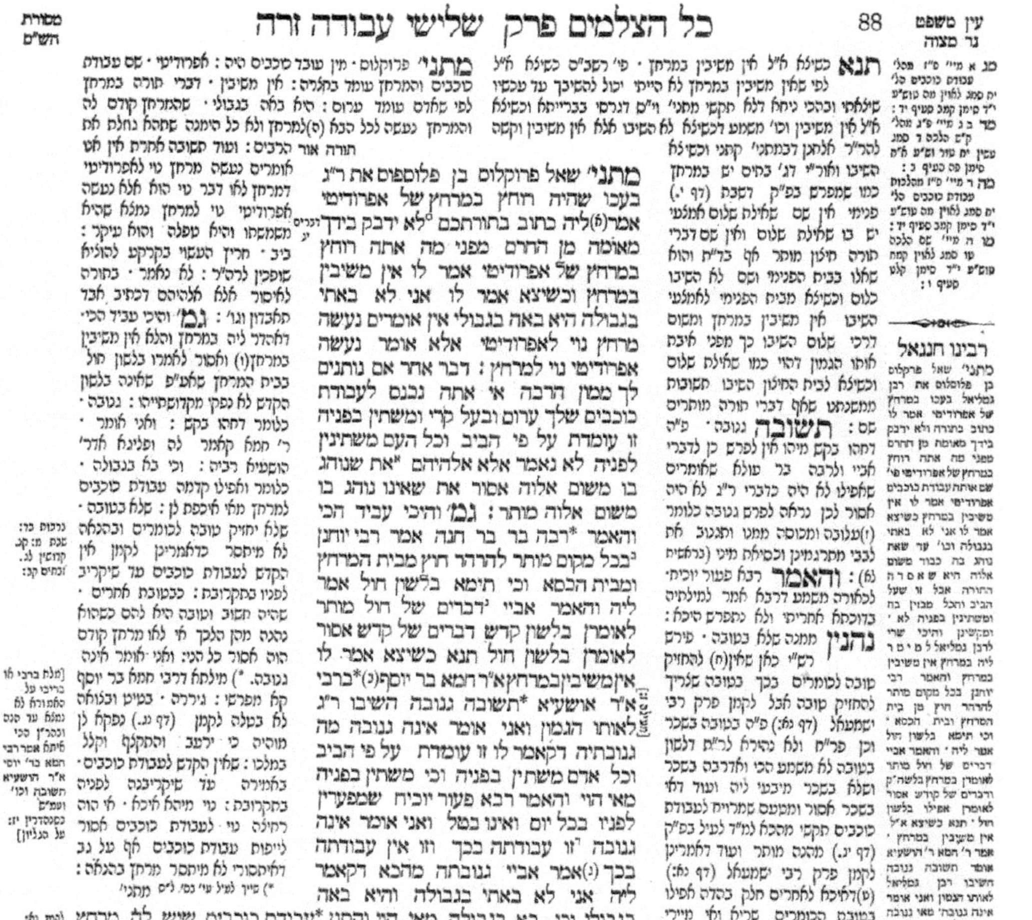

Figure 1: One of the 5,894 pages of the Babylonian Talmud.

When I was teaching at Columbia University, technoprophet Marshall McLuhan[26] came down from Toronto to lecture there. He talked about how the linear pattern of information resulting from print technology limited the thought patterns of people who learned from printed books. Word follows word, line follows line, paragraph follows paragraph, page follows page, chapter follows chapter, in a single necessary order from the first page to last. Learning through a medium that is a one-way street prevented creative, flexible, associative, open-ended, multidirectional, and multidimensional thought. Instead of just being authoritative, books became authoritarian, demanding thinking in straight lines from a fixed point of view. The book medium became a stronger message than its content. Designed to be read in privacy, in seclusion from others, the book ended dialogue. It conferred the values of isolation, detachment, passivity, and non-involvement.

I invited McLuhan to my office to show him how the Hebraic dialogic mindset, which could not tolerate unidirectional thought, used print technology to design multilinear books. I took a volume of the Talmud off my shelf and showed him non-linear pages designed in 16th century Venice. I opened it to page 2 (there is no page 1) and pointed to the patch of text in the center of the page that starts with the *Mishnah*, written in Hebrew, followed by the *Gemara*, in Aramaic. On one side is a column of Rashi's commentary in a different font than the central text. On the other side is a column of Tosafot followed by references to Rashi in Tosafot. In a narrow fourth column next to Tosafot, stacked vertically, are four different commentaries on commentaries that span centuries of dialogue over time and space. Sometimes, explanatory diagrams are printed on the side. In the margins around the page are numerous references to biblical passages and to other books spanning centuries. In new editions of the Talmud, commentaries and references extend into our postdigital age.

In the words of Rabbi Adin Steinsaltz: "The Talmud is thus the recorded dialogue of generations of scholars. It has all the characteristics of a living dialogue. Freshness, vivid spontaneity, and acute awareness of every subject permeate every argument and discussion. The spirit of life breathes on every single page."[27] It is not a set of books to be read in quiet solitude. We give life and continuity to the dialogue that began millennia ago by engaging the hundreds of voices talking across the folio pages in active dialogue with a learning partner. The two learners, a *ḫevrutah*, enter a page and move around within it while arguing with each other and calling for support from all the scholars before them. They can begin their learning on any of its 5,894 pages. The multivolume Talmud has no beginning and no end. The *ḫevrutah* can jump around within a page, between pages, between different Talmud tractates, look into the Bible, kabbalistic texts, or any other sources. A study hall in a yeshiva filled with many learning teams is a busy, dynamic, noisy environment, quite different from the eerie silence of a library for linear books.

When I began surfing the World Wide Web, it seemed a familiar place to me. I felt I had been there before. Talmud study had prepared me for its vast multidirectional options, hyperlinking and its non-sequential organization. I felt at home seeing home pages that had an uncanny resemblance to Talmud pages. As a member of the panel, "Toward an Aesthetic for the 21st Century: Networking, Hypermedia, and Planetary Creativity," at the 1990 conference of the College Art Association, I explored this confluence between traditional Jewish media experiences and encountering the emerging Internet.[28] A decade later, Jonathan Rosen wrote in *The Talmud and the Internet: A Journey between Worlds*:

> I can't help feeling that in certain respects the Internet has a lot in common with the Talmud. The Rabbis referred to the Talmud as a *yam*, a sea – and though one is hardly intended to "surf" the Talmud, something more than oceanic metaphors links the two verbal universes. Vastness and an uncategorizable nature are in part what define them both […] The Hebrew word for tractate is *masechet*, which means, literally, "webbing." As with the World Wide Web, only the metaphor of the loom, ancient and inclusive, captures the reach and the randomness, the infinite interconnectedness of words […] I take comfort in thinking that a modern technological medium echoes an ancient one.[29]

Canadian professor Eliezer Segal goes one step further. He uses the new medium to explicate the old. He created an interactive image-map site of a Talmud page on his website[30] to serve as a port of departure on a voyage through centuries of vital dialogue. The visitor to his site can click on any portion of the Talmud page image and be linked to a description of that patch of text. He explains the contents and purposes of the text in English, also describing when and where that patch of text was composed. The site visitor joins a community of explorers who weave learning through time and space.

There are groups of learners worldwide who study one page of the Talmud together every day completing the entire Talmud in seven years. To mark the end of the seven-year cycle and the beginning of the next cycle, the many *daf yomi* (a page a day) learning groups throughout the New York area pack Madison Square Garden in celebration. Unlike a book by a single author read alone in silence, the Talmud is a collaborative enterprise that creates community and continuity. Wiki tribes of Internet *daf yomi* learners are proliferating. My Google search for "daf yomi" for the first edition in 2006 yielded 104,000 sites and my Yahoo search yielded 81,100 sites. In 2010, there are 375,000 Google "daf yomi" sites and 812,000 Yahoo sites. In English we say, "he's an educat*ed* man," in the past tense. The Hebrew equivalent, *talmid ḫakham*, means "wise learner," one whose learning is daily and lifelong.

In the online magazine, *Computer-Mediated Communication*, Rensselaer Polytechnic Institute professor David Porush writes that the Talmud is an early example of hypertext.

> A page of Talmud is structured around a single text surrounded by concentric layers of commentary and commentary on commentary. By form and content, it announces the unfinished quality of constructing knowledge and the collective construction of shared values. Even in its layout on the page, the Talmud suggests a kind of time and space destroying hypertextual symposium rather than an authoritative, linear, and coherent pronouncement with a beginning and ending written by a solitary author who owns the words therein […] The notion of private self, or the notion of singular origin of knowledge,

> pales into insignificance in the face of this talmudic-hypertextual-Internet-like vision of communally-constructed knowledge.[31]

The living dialogue of the Talmud can provide clues for the creation of dialogic art. It can propose new creative options for making webart, blogart and wikiart that simultaneously address postmodern aesthetic and spiritual issues. It can be art that moves from deconstruction to reconstruction and from alienation to caring community.

Engaging the Bible in a playful spirit

Throughout this book, I engage the Bible in creative play through both my conceptual and artistic explorations. Lest the reader views my playing with the Holy Scriptures as irreverent and sacrilegious, I show how the Bible itself teaches us to approach it in a playful spirit. In *Psalm 119:174*, we read: "Your Torah is my plaything (*sha'ashua*)." *Sha'ashua* is a toy to engage children in play. In *Proverbs 8:30, 31*, King Solomon speaks in the voice of the Torah: "I [the Torah] was the artist's plan. I was His [God's] delight every day, playing before Him at all times, playing in the inhabited areas of His earth, my delights are with human beings." This translation from the Hebrew original is based on the ancient wisdom on the first page of *Midrash Rabba*.[32] God as the master artist played creatively with the Torah, His plan for creating the universe. *Midrash Rabba* uses these two verses from *Proverbs* to explain the first words of the Bible, "In the beginning God created." God first created "Beginning" referring to the Torah as an open-ended blueprint for creating the world. We learn this from an earlier verse, *Proverbs 8:22*, "God made me [the Torah] as the beginning of His way, before His deeds of yore." In human emulation of God's delight, we are invited to play with the Torah as we create new worlds.

Rabbi Abraham Yitzhak Kook, a down-to-earth mystic who served as Chief Rabbi of the Land of Israel, wrote a letter of congratulations on the founding of the Betzalel Art School in Jerusalem. By way of allegory, he refers to the revival of Hebraic art and aesthetics after two thousand years of exile as a child in a coma who awakes calling for her doll.

> The pleasant and beloved child, the delightful daughter, after a long and forlorn illness, with a face as pallid as plaster, bluish lips, fever burning like a fiery furnace, and convulsive shaking and trembling, behold! She has opened her eyes and her tightly sealed lips, her little hands move with renewed life, her thin pure fingers wander hither and thither, seeking their purpose; her lips move and almost revert to their normal color, and as if through a medium a voice is heard: "Mother, Mother, the doll, give me the doll, the dear doll, which I have not seen for so long." A voice of mirth and a voice of gladness, all are joyous, the father, the mother, the brothers and sisters, even the elderly man and woman who, because of their many years, have forgotten their children's games.[33]

Rabbi Kook saw artists at work as a clear sign of the rebirth of the Jewish people in its homeland. Their playful spirit nurturing sensitivity for beauty "will uplift depressed souls, giving them a clear and illuminating view of the beauty of life, nature, and work."

To give the reader a taste of making a biblical text a plaything, I will describe a traditional encounter with Jacob's ladder. "He had a vision in a dream. A ladder was standing on the ground, its top reaching up towards heaven as divine angels were going up and down on it" (*Genesis 28:12*). Learning Torah playfully is like tasting sweet fruits that we pick and eat while wandering through a grove. Therefore, *PaRDeS*, the word for grove or orchard, becomes an acronym for four levels of playing with the Torah. ***P'shat*** is the simple, literal meaning of the biblical words. ***Remez*** is a hint of innate significance. ***Drash*** is a homiletic interpretation. And ***Sod*** is a hidden, mystical meaning.

That a ladder is a ladder is the *p'shat*. That the ladder was spiral, like a spiral staircase, is the *remez*. We arrive at the spiral shape of the ladder by noticing that the numerical value of the Hebrew words for "ladder" and for "spiral" are both 130. Creative play using numerical equivalents of Hebrew letters, a system called *gematriah*, can lead to fresh insights. That the ladder was a metaphor for Mount Sinai with its top reaching towards heaven from the ground below is *drash*. The numerical value of *Sinai* is also 130. The deepest significance of the ladder as symbolized in *sod* is Jacob's body as well as the body of every human being whose head in the clouds can dream of what can be while his feet rest on the ground where dreams can be realized.

"Play in earnest" could be shared as the motto for Torah study and for postmodern art and science. Scientists are joining artists in recognizing and honoring the playful spirit in their creative work. Parallel to Rabbi Kook's use of a child's doll as a metaphor for art, J. Bronowski in his book *Science and Human Values* relates artistic activities and scientific research to the play of children and young animals.

> In science and in the arts the sense of freedom which the creative man feels in his work derives from what I call the poetic element: the uninhibited activity of exploring the medium for its own sake, and discovering as if in play what can be done with it. The word play is in place here, for the play of young animals is of this kind – an undirected adventure in which they nose into and fill out their own abilities, free from the later compulsions of need and environment. Man plays and learns for a long time (he has a longer childhood) and he goes on playing into adult life: in this sense of free discovery, pure science is (like art) a form of play.[34]

In a playful spirit, I weave throughout the book my insider experiences as a creative artist to clarify and exemplify how Hebraic consciousness can inform postmodern art through imagery concepts and visual *midrash* (creative biblical commentary). My artwork explores relationships between postmodern art in the postdigital age and Jewish patterns of thought, collaborative art and community participation, and space-time systems and electronic technologies. Intergenerational collaboration rather than solitary production and active participation rather than passive observation exemplifies the paradigm shift from Hellenistic to Hebraic consciousness.

Notes

1. Thorleif Boman, *Hebrew Thought Compared with Greek* (New York: Norton, 1960), p. 27.
2. Winston Churchill, *History of the Second World* War, vol. V (Boston: Houghton Mifflin, 1951), p. 532.
3. Bruno Zevi, "Hebraism and the concept of space-time in art," in Andrea Oppenheimer Dean, *Bruno Zevi on Modern Architecture* (New York: Rizzoli, 1983), 165.
4. Norris Kelly Smith, *Frank Lloyd Wright: A Study in Architectural Content* (New Jersey: Prentice-Hall, 1966).
5. J. Boyd, *The Papers of Thomas Jefferson*, vol. I. 1760–1776. (Princeton: Princeton University Press, 1950) and *The Great Seal of the United States* (Washington: US Department of State, Bureau of Public Affairs, 1966).
6. Alfred J. Kolatch, *Great Jewish Quotations* (Middle Village, NY: Jonathan David, 1996), p. 472.
7. A. Efland, K. Freeman, and P. Stuhr, *Postmodern Art Education* (Reston, VA: National Art Education Association, 1996).
8. Bruno Zevi, "Hebraism and the concept of space-time in art," 165.
9. Francesco Dal Co and Kurt Foster, *Frank O. Gehry: The Complete Works* (New York: Monacelli Press, 1998), p. 29.
10. Ibid. p. 30.
11. Ilya Prigogione and Isabelle Stengers, *Order Out of Chaos: Man's Dialogue with Nature* (New York: Bantam Books, 1984).
12. Peter Weibel and Timothy Druckery (eds.), *net_condition: Art and Global Media* (Cambridge, MA: MIT Press and Karlsruhe, Germany: ZKM Center for Art and Media, 1999), p. 19.
13. Wassily Kandinsky, *Concerning the Spiritual in Art*, trans. M. T. H. Sadler (New York: Dover, 1977), p. 1, 4.
14. Menachem M. Schneerson, *Torah Studies*, adapted by Jonathan Sacks (London: Lubavitch Foundation, 1986), p. 320, original italics.
15. Joseph B. Soloveitchik, *Halakhic Man*, trans. Lawrence Kaplan (Philadelphia: Jewish Publication Society of America, 1983), p. 92, 94.
16. E. L. Doctorow, *City of God* (New York: Plume Book/Penguin Putnam, 2001), p. 254.
17. Arthur David Ritchie, *Studies in the History and Methods of Science* (Edinburgh: The University Press, 1958).
18. Mel Alexenberg, "On the Creation of LightsOROT: Spiritual Dimensions of the Electronic Age," in *LightsOROT*, Mel Alexenberg and Otto Piene (eds.) (Cambridge, MA: MIT Center for Advanced Visual Studies, and New York: Yeshiva University Museum, 1988), p. 55.
19. Tsion Avital, *Art Versus Noart: Art Out of Mind* (Cambridge, UK: Cambridge University Press, 2003), p. 403.
20. Eric Mark Kramer, "Gebser and Culture," in *Consciousness and Culture*, E. M. Kramer (ed.) (Westport, CT: Greenwood Press, 1992).
21. Adin Steinsaltz, "The Imagery Concept in Jewish Thought," *The Strife of the Spirit* (Northvale, NJ: Jason Aronson, 1988), pp. 63–70.
22. David Gelernter, *The Muse and the Machine: Computerizing the Poetry of Human Thought* (New York: The Free Press, 1994).
23. David Gelernter, *Judaism: A Way of Being* (New Haven and London: Yale University Press, 2009).
24. David Gelernter, "Judaism Beyond Words," *Commentary*, May 2002, p. 32.

25. Gelernter, *Judaism: A Way of Being*, p. 174.
26. See Marshall McLuhan, *Guttenberg Galaxy: The Making of Typographic Man* (New York: New American Library, 1969) and *Understanding Media* (Cambridge, MA: MIT Press, 1994).
27. Adin Steinsaltz, *The Talmud: The Steinsaltz Edition*, Reference Guide (New York: Random House, 1989).
28. For a summary of this panel discussion, see: Ray Gallon, "Ideas of Planetary Creativity: The College Art Association Tackles Networking," *Netweaver*, http://cgi.gjhost.com/~cgi/mt/netweaverarchive/ideas_of_planetary_creativity_1290.html, Accessed February 24, 1996.
29. Jonathan Rosen, *The Talmud and the Internet: A Journey between Worlds* (New York: Farrar, Straus and Giroux, 2000), p. 7, 8, 11.
30. Eliezer Segal, "A Page for the Babylonian Talmud," http://people.ucalgary.ca/~elsegal/TalmudPage.html, Accessed October 8, 2010.
31. David Porush, "Ubiquitous Computing vs. Radical Privacy: A Reconsideration of the Future," *Computer-Mediated Communications Magazine*, vol. 2, no. 3, March 1, 1995, http://www.ibiblio.org/cmc/ mag/1995/mar/last.html, p. 46.
32. *Midrash Rabbah: Genesis*, trans. H. Freeman and Maurice Simon (London: Soncino Press, 1961), p. 1. *Midrash Rabba* are biblical commentaries collected over a period of 1,500 years and edited in the 6th century.
33. Abraham Yitzhak Hakohen Kook, *Rav A. Y. Kook Selected Letters*, trans. Tzvi Feldman (Ma'aleh Adumin, Israel: Ma'aliot Publications of Yeshivat Birkat Moshe, 1986), p. 191.
34. J. Bronowski, *Science and Human Values* (New York: Harper & Row, 1965), p. 76.

Chapter 1

Postdigital Perspectives: *Rediscovering Ten Fingers*

The digital moment has passed.[1]

Roy Ascott

Digital technology is significant, but it is part of something much more momentous.[2]

Stephen Wilson

Like air and drinking water, being digital will be noticed only in its absence, not by its presence. Face it – the Digital Revolution is over.[3]

Nicholas Negroponte

Postdigital is a term which has recently come into use in the discourse of digital artistic practice. This term points significantly to our rapidly changed and changing relationships with digital technologies and art forms. It points to an attitude that is more concerned with being human than with being digital.[4]

Wikipedia

Postdigital perspectives emerging from the paradigm shift from Hellenistic to Hebraic consciousness are reflected in the very word *art* itself. *Art* in English and French, *arte* in Spanish, *l'arte* in Italian, *Kunst* in German and Dutch, *iskustvo* in Russian, as in all European languages are related to artificial, artifact, imitation and phony. These words express the ancient Greek philosopher Pliny's praise of the artist who painted grapes so realistically that they could fool the pigeons who would try to eat them. The meaning of the Hebrew word for art is not only different, it is the exact opposite. In Hebrew, artist (*oman*) is spelled with the same letters (*AMN*) as the word *amen* which means truth. Its feminine form is *emuna*, faith, and as a verb *l'amen* means to educate. Postdigital artists in our networked world are assuming roles of creative educators who invite us to collaborate in creating lively artforms that integrate digital technologies with myriad aspects of our everyday life. They create participatory artworks to encourage us to learn from one another, sharing our unique ways of knowing, and in the words of the Hasidic master Rabbi Schneerson creating "a welcome environment for the interaction of our souls, our hearts, our visions."[5]

In contrast to honoring the artist who imitates the creation, Judaism values the artist who imitates the Creator. Engaging in the creative process to bring into being something new is a central value of Judaism that we learn from the biblical creation story that culminates in human beings being created in the divine image and likeness (*Genesis 1:26*). We exhibit our being in the image and likeness of the Creator by expressing our creativity. Unlike the Hellenistic view of nature as a finished product, the Hebraic perspective views us as potential partners of God in the continuing process of creation. The significance of our active collaboration with God in making the world a better place is expressed in a tale told two millennia ago in *Midrash Tankhumah.* The Roman governor over the Land of Israel asked Rabbi Akiva, "Which are greater and more beautiful, man's creations or God's?" The governor was taken aback by the rabbi's response that human creation is more exalted than divine creation. As the Roman was questioning the rabbi's unexpected response, Rabbi Akiva served a plate of wheat grains to the Roman and took cakes for himself. The puzzled Roman asked, "Why do you take cakes for yourself while you give me raw grains of wheat?" The rabbi responded, "You prefer God's creation. I prefer man's."

Modern art moved art beyond imitating nature. It created art movements throughout the 20th century that undermined the Hellenistic definition of art. This deconstruction of the Greek ideal invites 21st century redefinitions of art for a networked world in a postdigital era. Postmodernism expanded the definition of art in directions that artist Allan Kaprow calls "the blurring of art and life"[6] where art moves closer to the Jewish concept of art striving for truth, faith and education. We learn from *Exodus 35:30* that the prototypic biblical artists, Betzalel and Oholiav, were endowed with the talent to teach others to collaborate in creating a Tabernacle made from curtains and Lego-like parts. This modest tent-like structure was continually deconstructed and reconstructed during the Israelites' trek through the Sinai desert. Like the artists who were simultaneously educators in biblical times, in our postdigital age of participatory art forms artists are once again assuming pedagogical roles. Most postdigital artists hold teaching positions in art schools and university art departments and are authors of academic papers and books.

The author of *The Art Teacher's Guide to the Internet*, University of Florida art education professor Craig Roland, created a blog to continually update his book with innovative new ideas. I noticed a May 2009 posting "Pausing for a post-digital moment" in which he explained why he changed the tile of his blog from *The Art Teacher's Guide to the Internet: Ideas, Tools, and Resources for Teaching Art and Design in a Digital Age* to *in a Post-Digital Age.*

> I believe that in a broad sense adopting a "post-digital" mindset means, as Negroponte suggested over a decade ago, moving beyond teaching about technology itself to focusing on what we can do with it – the new opportunities that it provides for creative expression, visual communication and learning [...] In recent years, we've witnessed the emergence of new ways to experience the World Wide Web. The term "Web 2.0" has been coined to describe the Web's transition from a collection of static Web sites containing information to a more dynamic, interactive, social, and content-sharing environment. Web 2.0 tools and services like blogs, wikis, podcasts, photo- and video-sharing sites, social networks, and virtual worlds are now being used daily by millions of people around the globe to connect, communicate, collaborate, create, care, and share with others.[7]

A Google search identifies the first exhibition using "postdigital" in its title as the 2008 Polk Museum of Art's *Digital Art in a Post-Digital Age: Works from Florida Faculty*, a juried competition open to all art instructors and professors at the colleges and universities throughout the state of Florida. Other shows that have followed Polk's lead are the Tampa Museum of Art's exhibition *Bit, Byte, Dot, Spot: Postdigital Art*; *Kiosk: Artifacts of a Post-Digital Age* held at the STRP Art & Tech Festival in Eindhoven, The Netherlands; and *Code, Interactivity and Network* in 2010 at the Victoria and Albert Museum in London. There was also a *Post Digital Fury* symposium held in Tromsø, Norway, in which artists exploring postdigital issues in their artwork discussed the problems related to media art in an age when the digital hype has disappeared and media technology is associated with the noisy backdrop of our contemporary culture.

In addition to learning from *Wikipedia* that "postdigital" is a term used in the discourse of digital artistic practice at a time of rapidly changing digital technologies and art forms that are more concerned with being human than with being digital, we also learn: "The word *digital* comes from the same source as the word digit and *digitus* (the Latin word for *finger*), as fingers are used for discrete counting." The digital age reduced ten fingers to two. It replaced tactile counting with ten fingers with the mind game of binary counting – one as 01, two as 10, three as 11, four as 100, five as 101, six as 110, seven as 111, eight as 1000, nine as 1001, and ten as 1010. The digital age opened up the high tech option of creating artworks with clean fingers confined to a qwerty board or laser mouse. Messy fingers immersed in wet clay or smelly paint were out. Postdigital art is emerging from a vital dialogue between high tech and high touch experiences. It invites the rediscovery of ten fingers by adding the human touch to digital technologies in ways that I outline in my definition of "postdigital" posted on *Wiktionary* and *Wikipedia.* My act of collaborating in the creation of the world's most actively used dictionary and encyclopedia exemplifies the postdigital perspective.

> **Postdigital** (adjective), of or pertaining to art forms that address the humanization of digital technologies through interplay between digital, biological, cultural, and spiritual systems, between cyberspace and real space, between embodied media and mixed reality in social and physical communication, between high tech and high touch experiences, between visual, haptic, auditory, and kinesthetic media experiences, between virtual and augmented reality, between roots and globalization, between autoethnography and community narrative, and between web-enabled peer-produced wikiart and artworks created with alternative media through participation, interaction, and collaboration in which the role of the artist is redefined.[8]

Robert Pepperell, multimedia artist and musician, and Michael Punt, film-maker and writer, collaborated in writing *The Postdigital Membrane: Imagination, Technology and Desire*[9] that proposes a new metaphor for a postdigital perspective that extends beyond the limitations of a binary model of reality. They use the term "postdigital" to acknowledge the current state of electronic technology while rejecting the implied conceptual shift of the digital revolution's on/off, zero/one logic. The intellectual restrictions of the digital paradigm must be rejected because it insists on the reduction of continuous reality into discrete binary units. The wonderful complexity, ambiguity, and unpredictability of human experiences cannot be reconciled in the binary codes of digital data processing. The metaphor of an organic membrane shows how things can be both separate and connected. A semi-permeable cell membrane is a living system that both contains the protoplasm and regulates the flow of molecules between the inside and outside of the cell. The membrane metaphor gives form to complex phenomena (such as imagination, technology, and human desire) that require a paradigm of a dynamic living continuum that is not fragmented into billions of sets of two.

The transition from a digital to a postdigital age is the subject of Daniel Pink's book *A Whole New Mind: Moving from the Information Age to the Conceptual Age* (paperback edition renamed *A Whole New Mind: Why Right-brainers will Rule the Future*). Pink provides extensive evidence for the emergence of a postdigital Conceptual Age beyond the digital Information Age. The Conceptual Age is an age of creators and empathizers who activate the right hemisphere of their brains to compliment the left-hemispheric dominance of the Information Age of knowledge workers and the earlier Industrial Age of factory workers. It is a postdigital age in which well-developed high tech abilities are no longer enough for succeeding. In the Conceptual Age, high tech knowledge is integrated with high concept and high touch creativity in all areas of human endeavor.

> High concept involves the ability to create artistic and emotional beauty, to detect patterns and opportunities, to craft a satisfying narrative, and to combine seemingly unrelated ideas into a novel invention. High touch involves the ability to empathize, to understand the subtleties of human interaction, to find joy in one's self and to elicit it in others, and to stretch beyond the quotidian in pursuit of purpose and meaning.[10]

Thomas Friedman identifies the forces that are determining the nature of a postdigital world in his book *The World Is Flat: A Brief History of the Twenty-First Century.*[11] The flattening of the world is creating a level playing field in which Israel with a population of only 7 million can successfully compete with giants such as the United States, Europe, India and China. He labels these flattening forces as the new age of creativity: when the walls came down and the Windows went up; the new age of connectivity: when the Web went around and Netscape went public; and the new age of collaboration: harnessing the power of communities. Those forces of creativity, connectivity, and collaboration are shaping postdigital art.

I invited some of the world's most innovative thinkers on postdigital futures – artists, art educators, and writers on new media arts – to contribute chapters to my book *Educating Artists for the Future: Learning at the Intersections of Art, Science, Technology, and Culture.*[12] Stephen Wilson, author of *Information Arts,*[13] proposes that although the impact of digital technology is significant, it forms part of something much more momentous that is intertwined with the aesthetic, ethical, cultural, and socio-economic. Art is being redefined by a digital revolution linked to revolutions brewing in the realms of biology, neurophysiology, materials science, and cosmology that call for new directions. Roy Ascott, author of *Telematic Embrace*,[14] coins the word *moistmedia* as the symbiosis between dry pixels and wet biomolecules. Postdigital art will be devoted to creating moistmedia artworks from which new metaphors, new language, and new methodologies will arise. The dynamic interplay between digital, biological, and cultural systems calls for a syncretic approach realized through connectivity, immersion, interaction, transformation and emergence. Stefan Sonvilla-Weiss, author of *(In)Visible: Learning to Act in the Metaverse*,[15] sees the postdigital artist as an intermediary, a catalyst between diverse fields of knowledge, ways of thinking, social models, and solution strategies based upon cooperation, communication and interaction. Network culture not only changes modes of media production and distribution, but it transforms art from object making to art as processes of creating immaterial, rhizome-like structures of remotely connected individuals in online communities. "Print and radio tell; stage and film show; cyberspace embodies."

The artworks of three other contributors to *Educating Artists for the Future* represent a range of postdigital directions and their international flavor. Michael Bielicky from the Czech Republic is a professor of infoart/digital media at the Hochschule fur Gestaltung/ZKM Center for Art and Media in Karlsruhe, Germany; Yacov Sharir from Israel is a professor of dance and multimedia art at the University of Texas at Austin; and Eduardo Kac from Brazil is a professor of art and technology at the School of the Art Institute of Chicago and author of *Signs of Life: Bio Art and Beyond*.[16]

Bielicky created *This Year in Jerusalem*, an interactive installation at the Robert Guttmann Gallery of the Jewish Museum in Prague that telematically transported gallery visitors to Jerusalem. Bielicky used a motion capture tracking system to record emitted infrared rays reflected off the bodies of visitors to the gallery to generate data for creating a 3D avatar of the visitor. As the visitor moved in the gallery space, his moving avatar was transported into a full-wall projection of the Western Wall in Jerusalem streaming in real time from a web camera beaming via a high-speed Internet connection into the gallery. Gallery visitors in Prague experienced their virtual selves interacting with people in Jerusalem without having to overcome any geographical distance between the two cities. Bielicky created *Falling Times* in collaboration with his wife Kamilla B. Richter. It is a series of artworks driven by the Internet news headlines that are translated into a universally understood dynamic pictogram language projected in public spaces. They see postdigital art as breaking out of institutionalized art frameworks to reach out to a wide public that does not frequent galleries or museums. Equipped with a car, a laptop, a compact powerful projector, and a small power generator, they create an instant presence in the urban landscape by covering the facades of big buildings with news-driven falling icons that transform static city architecture into lively animated organisms. In their most recent *Data Driven Narratives*, they extend the non-stop generation of web news to changing weather patterns and the ups and downs of the stock exchange, data flows programmed to provide alternative options in shaping storytelling narratives.

Sharir's artistic quest that led to the immersion of his wired body in virtual environments had its origins in his childhood studies with a violin teacher whose way of playing made him feel as if his teacher was wearing the music, as if the music had become his second skin. He taught young Yacov that technological success playing the right notes was not enough without learning to embody the mood, the atmosphere, the colors, and the appreciation of meaning. "Tell me your story through your music," his violin teacher insisted. He studied art and dance in Jerusalem, he lived two professional lives simultaneously – shaping raw earth into works of clay art while transforming his body into art material as a dancer with the renowned Bat Sheva Dance Company. He was invited to teach at the University of Texas where he founded the American Deaf Dance Company which pioneered the inclusion of deaf artists in professional dance and the Sharir Dance Company for which he choreographed and created original technologically mediated compositions using wearable computers, technologically charged interactive systems, virtual reality, interactive video art, and computer animated cyber-human performers. He teaches his students how to build content and meaning through the collaborative creation of technologically mediated interactive art that explores how the disembodied digital self is re-embodied in cyber-bodies occupying increasingly immersive cyber-worlds.[17]

The artwork of Eduardo Kac exploring the boundaries between humans, animals, and robots exemplifies the postdigital movement integrating digital with biological systems. As an artist, Kac is "interested in reflecting on the multiple social implications of genetics, from unacceptable abuse to its hopeful promises, from the notion of 'code' to the question of translation, from the synthesis of genes to the process of mutation, from the metaphors employed by biotechnology to the fetishization of genes and proteins, from simple reductive narratives to complex views that account for environmental influences."[18] Kac's 2010 exhibition *Lagoglyphs, Biotopes and Transgenic Works* at Oi Futuro in Rio de Janeiro surveys the wide range of his artworks. The *Lagoglyphs* are a series of artworks in different media in which he creates a visual language and form of writing that he describes as "rabbitographic." The series references Kac's project *GFP Bunny* – also known as "Alba, the green bunny" – a genetically modified rabbit that glows fluorescent green under certain light conditions. The "lagoglyphs" are pictograms, visual symbols composed of two units (one green, one black) that each represent the bunny and resist any assigned meaning. *Lagoglyphs: The Lepus Constellation Suite* explores the notion that certain life forms, such as a green rabbit, are "alien" to us. The project literally communicates a sign system to the "alien" other. The lagoglyphic "message" inscribed on each disk has been transmitted to the Lepus Constellation (the Gamma Leporis star below Orion) from Cape Canaveral, Florida, on March 13, 2009 via satellite broadcasting equipment and a parabolic dish antenna. Kac quite literally connects sign systems across the universe: the lagoglyphic rabbit meets the lepus (Latin for hare) constellation. The constellation is 29 light years away from Earth, which means that Kac's messages would arrive in the year 2038. *Lagoogleglyph* is a distributed, global artwork that inscribes lagoglyphs into the environment and makes them visible to the world. In its first manifestation *Lagoogleglyph* consists of a pixelated lagoglyph, referencing a rabbit head, inscribed on the roof of the Oi Futuro building and specifically made by Kac for the eye of a satellite used by Google. The artist hired the same satellite used by Google and produced a photograph identical to the kind used by Google Earth. *Lagoogleglyph* is potentially visible to anyone on the planet via Google's geographic search engine.[19]

Eduardo Kac has been a pioneer of transgenic art, a new artform based on the expression of synthetic genes in a host organism, or the expression of an existing gene in a different organism. *Cypher* is a fusion of a sculptural artist's book and a do-it-yourself transgenic kit containing a minilab with Petri dishes, nutrients, and synthetic DNA, which has a poem written by Kac embedded in its genetic sequence. At the center of the *Natural History of the Enigma* series is Edunia, a genetically engineered flower – a hybrid of Kac's DNA and a petunia flower – that expresses Kac's DNA in its red veins. *Specimen of Secrecy about Marvelous Discoveries* is a series of works that Kac describes as "biotopes," art objects that are living organisms. The biotopes are ecologies of thousands of micro-organisms framed in a transparent casing with earth, water, and other materials. Together, the individual life forms constitute an artwork that is an ecology and living system, with behaviors that express themselves in a visual slow-motion animation. Kac orchestrates the biotopes' metabolism through combining them with nutrient-rich media and controlling the energy they receive in order to keep them alive. Colors and geometric visual forms are responses to the external environment, such as temperature and light, and the internal metabolism of the biotope.[20]

Roots and globalization

The most powerful forces shaping the postmodern condition are globalization and digital technologies, free trade and the free flow of information, and the range of human reactions to them, from high tech to high touch responses. The roots of globalization and its links to new technologies and art can be traced to Rembrandt's Netherlands. Free-market capitalism was invented in this first European republic during the 17th century. The Dutch formed the first public corporation (the East India Company), the first stock exchange (the Amsterdam Bourse), and the first commodity exchange in which tulip bulb futures were traded. The merchants and farmers who pooled their resources to form the first public corporations made themselves wealthy enough to buy paintings for their homes. Artists flooded the market with portable art that could be easily traded by using the new art-making technologies of oil painting on canvas and etchings printed on paper. This lightweight merchandise available for purchase on the free market was quite different from the permanently installed fresco or tempera paintings that were commissioned for the walls of palaces or churches.

In her book, *Rembrandt's Enterprise: The Studio and the Market*,[21] Svetlana Alpers presents a fascinating study of Rembrandt the artist as entrepreneur and businessman. His successes as a player in the art marketplace is what brought him honor in contrast to artists who sought honor through the patronage of the aristocracy or church. A brisk trade in Rembrandt's promissory notes paralleled the activities of the stock and commodity markets of Amsterdam.

When people who had paid him for a future painting got tired of waiting for him to make it, they sold their promissory note to another admirer of Rembrandt's work with more patience. In addition to painting, he made etchings that could be printed in multiples and sold to a mass market. When he would sell out one edition, he would rework the etching plate and print a new revised edition. He could often be seen at art auctions bidding for his own etchings to drive their price up. Rembrandt's enterprise represents the values established in the marketplace and of the individual suited to that market. These two reciprocal values constitute the ideology of the free market and the free individual that contemporary democracies have inherited in great part from the example of Rembrandt's Holland.

Thomas Friedman argues in *The Lexus and the Olive Tree: Understanding Globalization*: "The challenge in this era of globalization – for countries and individuals – is to find a healthy balance between preserving a sense of identity, home and community and doing what it takes to survive within the globalization system."[22] Friedman uses the Japanese luxury car Lexus as a metaphor for globalization in contrast to the centuries-old olive trees in the Land of Israel as the metaphor for the connection of a people to its roots. When he visited the Lexus factory in Toyota City, he watch in fascination as 310 robots produced 300 Lexus sedans each day while most of the 66 human beings who worked there were engaged in quality control. The Lexus represents the fundamental human drive for growth, improvement, and prosperity as it is played out in today's globalization system of free trade and free flow of information. Olive trees represent the security of being rooted in one's own culture to provide strength to cope with the onslaught of globalization and the rapid changes in its concomitant digital technologies.

> Olive trees are important. They represent everything that roots us, anchors us, identifies us and locates us in this world – whether it be belonging to a family, a community, a tribe, a nation, a religion or, most of all, a place called home. Olive trees are what give us the warmth of family, the joy of individuality, the intimacy of personal rituals, the depth of private relationships, as well as the confidence and security to reach out and encounter others [...] A country without healthy olive trees will never feel rooted or secure enough to open fully to the world and reach out into it. But a country that is only olive trees, that is only roots, and has no Lexus, will never go, or grow, very far. Keeping the two in balance is a constant struggle.[23]

The saga of the Jewish people from the time of Abraham 3,800 years ago to its contemporary return to its roots in its ancestral homeland after its global dispersion can be instructive in understanding globalization and its impact on art. Scholars of cultural change and visual culture use the Jewish Diaspora as the prototype of a global people.[24] A sweeping overview of the biblical narrative as it reaches into the 21st century can reveal how Jewish consciousness was shaped by the dynamic interplay between family and humanity, between particular and universal, between deep connections to their homeland while dispersed across the globe, and between pride in roots and reaching out to explore the cultures of others.

The Hebrew Bible is the most widely translated and read book. It communicates a universal message through telling the story of a single family evolving into the nation of Israel. It shows how a close look at one particular culture can shed light on human similarities. The biblical narrative begins with the creation of the universe and the trials and tribulations of the common ancestors of all humanity – Adam and Noah. The Tower of Babel project was the early version of globalization, a project of all the people of the world joining together for a common purpose. It resulted in disaster because it created a single homogenized culture that eliminated individual differences and cultural diversity. Today's inevitable globalization process can be equally disastrous if it fails to recognize and honor differences between families, tribes, religions, and nations.

In its third chapter, the Bible shifts its focus from all of humanity to the life of Abraham and the story of the Children of Israel. It begins with the divine command to leave one's familiar past in order to envision a new future. Abraham is told: "Walk yourself (*lekh lekhah*) away from your land, from your birthplace, and from your father's house, to the land that I will show you" (*Genesis 12:1*). A word *lekhah* "yourself" added to *lekh* "walk away" teaches that one can only come to see the new land by moving psychologically as well as physically away from an obsolete past. Abraham is identified as a Hebrew, literally "a boundary crosser." The personal power of Abraham to leave an obsolete past behind and to cross conceptual boundaries in creating a new worldview is a meaningful message for our postdigital age of globalization. He deserted the local gods of his father in which divine messages were perceived as flowing through the narrow channel of an idol's mouth. Instead, he gained the insight of the existence of an all-encompassing spiritual force that integrates the entire universe and beyond with all humanity in one universal ecosystem.

Abraham's son, Isaac, is the only one of the three patriarchs who spends his entire life in the Promised Land. He is the patriarch who roots his family in the land. Isaac's son, Jacob, however, uproots himself and goes alone to live in a foreign land. When he returns with his large family two decades later, he wrestles with an angel to free himself from his deceptive ways reinforced by his father-in-law. He is injured in his struggle and limps his way back to his roots with a new name, Israel (related to the word "straight") instead of Jacob (related the word "crooked").

In his old age, Israel leaves his land a second time to *Mitzrayim* ("narrow straits"). *Mitzrayim* is the biblical name for Egypt. Israel's family grows there in number as it becomes enslaved in the narrow perspective and alien ways of the totalitarian global power of the day. At the zero hour when all seems lost after centuries in *Mitzrayim*, the Israelites win their freedom and escape to the desert. Trekking through the desert while experiencing its wide-open expanses begins the process of leaving narrowness of thought behind and returning to the open-systems thought of their ancestor Abraham. Seven weeks later at the foot of Mount Sinai they are given the Torah, a blueprint for building a new life in freedom when they return to their land.

Leaders of the twelve Israelite tribes spy out the land from the wilderness of Tzin to Rehov, which can be translated as "wide expanses." The challenge was to abandon the narrowness of *Mitzrayim* and bring the expansive consciousness of the desert into every aspect of their lives in the villages and cities they would build in the Promised Land. Ten of the spies return to the desert encampment strongly opposing entering the Land. They were unable to escape their slave mentality and enter into the open-systems thought of a liberated people. Only Joshua and Calev meet the challenge. The Torah tells us that Calev of the tribe of Judah had "a different spirit." He was able to make the paradigm shift required to build a society in freedom. Unfortunately, the ten tribal leaders who were unable to make the shift wandered the desert for forty years and died there. The next generation born in the open desert rather than in the narrowness of *Mitzrayim* entered the Promised Land with Joshua and Calev.

After centuries struggling to realize the Torah blueprint free in their own land, seeming to be most successful under the leadership of David and his son Solomon, Jacob's family splits up into the kingdoms of Israel and Judea. The conquest of the northern kingdom of Israel and the forced dispersal of ten tribes led to their assimilation and disappearance. When Judea fell to the Romans, however, a plan of survival without national sovereignty was devised by the rabbis of Yavneh and codified later in the Talmud. It worked. Although the Jewish people (the tribes of Judah and Benjamin) from the kingdom of Judea were dispersed across the globe, they retained their Hebraic consciousness as "boundary crossers" for two millennia.

The 20th century ingathering of the Jewish people into their ancestral homeland of Israel at the time that many other peoples are being dispersed into new host countries would seem to be a countertrend to the powerful forces of globalization. However, the rebirth of the Jewish state and the ingathering of the exiles offer opportunities to plant roots that can provide the sure footing required to play the fast-moving globalization game. Six decades after its rebirth, Israel has become a major player in the global world of high tech. You are most likely using Israeli innovations in multiple aspects of your encounters with digital technology. These innovations result from a contemporary renewal of the ancient Israelite thinking that Boman characterizes as being "dynamic, vigorous, passionate, and sometimes quite explosive in kind."[25] Intel's Pentium 4 and Centrino microprocessors were entirely designed, developed, and produced in Israel. Your PC and laptop should be labeled "Israel Inside." Motorola, with its largest R&D center being in Israel, invented the cell phone. Most of the Windows NT operating system was developed at Microsoft's R&D center in Israel. Firewall security software, instant messaging (ICQ), ingestible video capsules to examine internal organs, are just a small part of a growing list of Israeli high tech creations.

In *Start-Up Nation: The Story of Israel's Economic Miracle*,[26] Dan Senor and Saul Singer explore how the creative and innovative thinking of "boundary crossers" rooted in a free society is exploding. Israel, barely the size of New Jersey with a smaller population, has more companies listed on the NASDAQ exchange than all companies from the entire European continent. The most significant measure of technological promise is venture capital. In 2008, per capita investments in Israel were 2.5 times greater than in the United States, more that 30 times greater than in all of Europe, 80 times greater than in China, and 350 times greater than in India. This amazing success can be attributed to the chutzpah and questioning attitude built into the Jewish consciousness over millennia coming to fruition in the national ethos of contemporary Israel.

Biblical commentators explain why Noah did not make it as the first Hebrew and only his descendent Abraham did. When God told Noah he should build an ark to save his family from a flood that would destroy the world, Noah followed God's orders without question. However, when God told Abraham that he was going to destroy the sinful city of Sodom, Abraham had the chutzpah to argue with God.

> He came forward and said, "Will You actually wipe out the innocent together with the guilty? Suppose there are fifty innocent people in the city. Would you destroy it, and not spare the place for the sake of the fifty good people inside it? It would be sacrilege even to ascribe such an act to You – to kill the innocent with the guilty, letting the righteous and the wicked fare alike. It would sacrilege to ascribe this to You! Shall the whole world's Judge not act justly?" God said, "If I find fifty innocent people in Sodom, I will spare the entire area for their sake." Abraham spoke up and said, "I have already said too much before my Lord! I am mere dust and ashes! But suppose there are five missing from the fifty innocent? Will You destroy the entire city because of the five?" God replied, "I will not destroy it if I find forty-five there." Abraham persisted and said, "Suppose there are forty there?" "I will not act for the sake of the forty." "Let not my Lord be angry, but I must speak up. What if there are thirty there?" "I will not act if I find thirty there." "I have already spoken too much now before my Lord! But what if twenty are found there?" "I will not destroy for the sake of the twenty." "Let my Lord not be angry, but I will speak just once more. Suppose ten are found there? "I will not destroy for the sake of ten." (*Genesis 18:23–32*)

Amos Oz, one of Israel's most talented writers explains that Judaism and Israel have always cultivated "a culture of doubt and argumentation, an open-ended game of interpretations, counter-interpretations, reinterpretations, opposing interpretations. From the very beginning of the existence of Jewish civilization, it was recognized by its argumentativeness."[27]

George Gilder, whose books have been honored as the most important works of the 20th century by *Wired* and *National Review*, explains in his 2009 book *The Israel Test*[28] how the tiny state of Israel is the pivotal point dividing the world into two fractious armies. The prime issue is not a global war of civilizations between the West and Islam or a split between Arabs and Jews, but a deeper moral and ideological conflict. He summarizes his Israel test with the questions: What is your attitude toward people who excel you? Do you aspire to their excellence, or do you seethe at it? Do you admire and celebrate exceptional achievement, or do you impugn it and seek to tear it down? So far in the 21st century, the Jewish three-tenths of 1 percent of the world population have been awarded 32 percent of the Nobel Prizes. Journalist Caroline Glick sums it up: "Some people admire success; some people envy it. The enviers hate Israel." The Israel test is a test of barbarism, envy, and death against civilization, creativity and life.

Jewish history is the prototype for the creative tension between envy and success, between subjugation and freedom, between exile and homecoming, between narrow unidirectional thought and open-ended systems thought as well as the energetic interplay between local action and global consciousness and between being rooted in one's own culture and exploring others. This tension and interplay is becoming the stimulus and raw material for forging fresh directions for art in our postdigital era.

Parentheses of Asia

Japan's greatest kite maker, Tsutomo Hiroi, came to fly his undulating giant dragon kite in the skies over Munich as part of the Sky Art '83 exhibition. He saw me building a *sukkah* at the entrance to the BMW Museum and offered his expertise in carpentry. I enthusiastically accepted. I explained to him that a *sukkah* is a fragile hut built to celebrate the holiday week of *Sukkot* to recall the Israelites' desert dwellings during their exodus from Egypt. It is sky art since it is built so that the stars in the night sky are visible through gaps in its roof. Jews throughout the world continue to build a *sukkah* every autumn to not only remember their past but to project a future of world peace. All humanity is invited to celebrate *Sukkot* to realize the biblical teaching of the prophet Zechariah (*14:16–19*) that if people everywhere would live for just one week in unprotected huts open to their neighbors and to the sky, then peace with each other and with the natural world would blossom.

As we worked, Hiroi chanted over and over again that the *sukkah* we were building was a beautiful Japanese building, a beautiful Japanese building. I kept insisting that it was a Jewish building. As we put the finishing touches on our *sukkah*, we finally compromised by agreeing that we had built an Asian building – Israel being on the west coast of Asia and Japan on the east coast. Its fragility and simple wooden construction spoke to both of our traditions. The following year, Hiroi invited me to lecture to his art students at the university in Tokyo where he was a professor. I brought sand I had scooped up from the beach in Tel Aviv on the Mediterranean coast to exchange with sand from the easternmost place in Japan, the fishing village of Chikura on the Pacific Ocean.

As we rode on a local train through rice paddies from Tokyo to Chikura, we continued the dialogue we had started during our *sukkah* building time in Germany. We explored common concepts of space and time in Jewish and Japanese cultural traditions and in ancient spiritual practices of Judaism, Shinto, Taoism and Buddhism. They reminded me of the discussions that I had had with Victor Solomon, my friend and neighbor in Teaneck during the years I taught at Columbia University. When I had moved with my family from Israel to New Jersey, Victor and his family had moved there from Japan where he had been chaplain at a US Air Force base. During his years in Japan as rabbi and air force colonel, he wrote a book on the Jews and Japanese that was translated into Japanese and became a best-seller.

When I returned to Israel after four years in Teaneck, I established a new college in the Negev desert. My discussions on the Jews and the Japanese continued in my talks with

Avigdor Shachan who would come down to my college as the head of the office in Israel's Ministry of Education responsible for all the colleges in Israel. He studied common threads revealed through the Hebrew and Japanese languages while tracing the migratory routes of the ten tribes exiled from the kingdom of Israel 2,700 years ago that led the tribe of Ephraim to Japan. We renewed our talks two decades later in his home facing wall-sized maps he had drawn to trace the migration of the exiled tribes. We were both teaching at Ariel University when he wrote about his research in his book *In the Footsteps of the Ten Lost Tribes.*[29] He leads tours to Japan that seek out the links between the ancient peoples from the west and east coasts of Asia including meeting with present-day Japanese B'nai Ephraim.

As the train pulled into the Chikura station, I realized that other traditions across Asia from Israel to India to China to Japan should be included in the conversation about the contemporary paradigm shift from the Hellenistic to the Hebraic roots of Western culture. Hiroi and I took a taxi from the town center to the coast where we found a beach of black volcanic sand. I found a stick with which I etched a parenthesis mark in the damp sand which I filled with the yellow sand I had brought from the Tel Aviv beach. I photographed one half of the parentheses of Asia as a black stripe below a turquoise stripe of the Pacific Ocean topped by a blue stripe of sky. Back home in Israel, I etched the other half of the parentheses in the yellow sand of the Tel Aviv beach and filled it with black sand I had brought from Chikura. I photographed the yellow stripe below a turquoise stripe of the Mediterranean Sea topped by a blue stripe of sky. At the Burston Graphic Center in Jerusalem, I made a serigraph *Parentheses of Asia* in which I screened both photographs set vertically. Israel's Mediterranean sky, surf, and sand are printed facing Japan's Pacific sky, surf, and sand. *Parentheses of Asia* is in the collections of the president of Israel and the emperor of Japan, an oceanographer.

As another Israeli artist casting her gaze to the Far East, Nurit Bar-Shai discovered the large number of webcams looking at the legendary Mount Fuji from different points of view, different resolutions, and different perspectives. She created *FUJI: Spaces and Other Places* that she considers to be a postdigital artwork. She appropriates Internet images from ten of the web-cameras combining them in a real-time composition of layered visuals that she streams live back to the Internet. The layered images in *FUJI* are not obvious and demand the viewer's involvement over time as the work unfolds from day to day, from month to month, from season to season during a period of a year. The frame of her artwork is not the webcam's border but the duration of the work. Her poetic gaze eastward to a distant place considered sacred by the Japanese is intriguing to her. Bar-Shai felt the creative urge to gaze from afar at the live online sublime beauty of a revered mountain at the other end of Asia. She employs available technology to reflect on the way we perceive distance, presence, and real-time experience through the Internet. Bar-Shai was inspired by earlier studies in which artists imply motion, change, and multiple points of view, from the artwork of 19th century Japanese artist Katsushika Hokusai's *36 Views of Mount Fuji* to Duchamp's *Nude Descending a Staircase*, Marey and Muybridge's photographic investigations of movement, and Hockney's photo collages.

When I was doing Internet research for my book *Educating Arts for the Future: Learning at the Intersections of Art, Science, Technology, and Culture*, I came across a MAAP (Multimedia Art Asia Pacific) conference in Singapore, "New Media Art, Technology and Education." A paper, "Self-organization, new media, and traditional Chinese philosophy," presented by Wengao Huang, a professor of media art at the College of Information Science and Engineering of Shandong University at Weihai, China, caught my eye. I tracked him down and asked him to send me a copy of his paper that seemed related to my work on new media and traditional Jewish philosophy. When it arrived I was amazed by the parallels between his explorations of art as the embodiment of Tao and my work on the confluence between ancient Jewish philosophy and postdigital art as expressed through *kabbalah*, a down-to-earth mystical tradition of Western civilization. Although we live and work at opposite ends of Asia – China and Israel – I had the opportunity to meet Wengao Huang in England at the 2006 Consciousness Reframed conference at the University of Plymouth and discuss our common background of being scientists turned artists exploring spiritual dimensions of the postdigital age derived from two different cultures. We were both fascinated that ancient spiritual traditions arising simultaneously on the east and west coasts of Asia millennia ago addressed creativity and the arts with a common language perhaps encoded in the shared consciousness of all humanity. Our dialogue prompted me to write "Ancient Schema and Technoetic Creativity"[30] in the journal of speculative research *Technoetic Arts* where I explore how schematic systems originating in Chinese and Jewish traditions offer fresh insights into the structure of human consciousness and its creative expression in technoetic art, art created at the intersections of art, science, technology, and consciousness research. Roy Ascott, organizer of the Conscious Reframed conference and editor of *Technoetic Arts*, describes technoetic art as art that arises from transdisciplinary research, collaboration, and innovation leading to new systems and structures that re-evaluate older worldviews and cultural practices in relation to artificial life, the promise of nanotechnology, the ecology of mixed media environments, the reach of telematic media, and the effect of a post-biological culture on human values and identity.

I invited Wengao Huang to write a chapter for my *Educating Artists for the Future* book. His chapter, "New Media Art as Embodiment of the Tao,"[31] presents evidence of how the natural sciences are creating a new paradigm that demonstrate what the ancient Chinese perceived as the Tao – dynamic monism in which matter is not concrete and the self not centered and unified. New media arts are embracing this holistic paradigm derived from complexity science that finds parallels in the traditional Chinese spirit's emphasis on connections, transformations and emergence. Generative and interactive art utilizing emerging technologies have great potential to vividly demonstrate the transforming spirit of the Tao and give new meanings to the saying within the traditional Chinese art world: "Art is the embodiment of the Tao." He describes his artwork on interactive morphogenesis that derives from his research in biotechnology, computer graphics, and ancient Chinese philosophy. The virtual creatures that he creates in artificial evolutionary developmental systems are based upon a biologically plausible developmental model guided by interactions between genes and a morphogenetic field of protein gradients. Artistic expression in postdigital art does not fit within boundaries of areas of conventional specialization. Underlying various art forms there is a unifying aesthetics based on *Ch'i*, which links the arts, humanity, and the universe in a great whole that promotes artistic creativity. Huang emphasizes that postdigital art is the alchemy of our time, bringing technology, consciousness, and culture into a global crucible that leads to the only reality – the Tao, transforming diversity within coherence.

In addition to Taoism, other ancient Asian spiritual systems such as Buddhism in Tibet and Korea and the Vedic traditions of India also offer innovative directions for the development of postdigital art forms. Their conceptual schemata show remarkable parallels to the kabbalistic model of creative process. When a group of Jewish leaders was invited by the Dalai Lama to visit him in northern India, they explored shared relationships between tantra and *kabbalah*. Roger Kamenetz documented their dialogue in his book *The Jew in the Lotus*.[32] After Rabbis Jonathan Omer-Man and Zalman Schachter mapped the kabbalistic system for the Buddhist master, "he leaned back and shook his head with amazement, smiling. He spoke rapidly in Tibetan to Lakor, who reported back his comment, 'It's interesting to find that there are very striking similarities.'" These similarities were also revealed in "Touching Light: Post-Traditional Immersion in Interactive Artistic Environments,"[33] a transcultural dialogue between Diane Gromala, a Canadian professor of interactive arts and technology and her Korean doctoral student, Jinsil Seo. Seo investigates the artistic significance of immersion in responsive and interactive spaces through her artworks. Her Asian cultural background colors her explorations of ambient immersive space, body interaction with environment, ephemeral and transient states of consciousness, and *Ch'i* as the flow of energy, in creating artworks in which bodily movements trigger interaction and immersion. Seo describes how her research and artmaking is influenced by growing up in South Korea where she developed a deep level of understanding of the conceptual connectivity between consciousness and immersion related to space and time derived from Buddhist thought.

In "Expressing with Grey Cells: Indian Perspectives on New Media Art,"[34] Vinod Vidwans, a professor of new media and software user interface design in India, presents an overview of a tradition of creativity in the arts that spans millennia on the Indian subcontinent. This overview forms the theoretical foundation for his description of the multifaceted confluence between traditional Indian perspectives and directions in art and education shaped by new media technologies. Art in both ancient India and the postdigital age share profound computational and mathematical foundations, convergence, and interdependence between fields, and an aesthetics of virtual reality. The Sanskrit word for art is *kala* – etymologically derived from the root that means counting, calculating or computation – indicates a deep level of correlation between artistic quantification and the quality of aesthetic experience. Art, science, and technology in India were integrated activities shaped by awe, curiosity, primordial quest, magic and mystery. The contemporary orchestration of art, science, and technology in a symphony of new media art mirrors the interdependent flow that guided artists in ancient India to integrate all realms of human creativity. Mathematical aesthetics coupled with the interdependence of various fields of human endeavor resonates powerfully in postdigital art forms in which the formal logic of algorithms offers infinite creative possibilities, ineffable corollaries of timelessness, and the unfolding of hidden metaphors in the unfathomable depths of structures of space and time.

Australian artist and writer Rob Harle wrote an insightful review of the first edition of this book for *Leonardo Reviews*.[35] His praise of the book was tempered by his main criticism that I should have looked for postmodernist directions in the art of Asian societies (underpinned by Buddhism) and of Chinese society (informed by Taoism) and not just have limited my research to exploring the similarities between postmodern art and the dynamic, conceptual attributes of the Torah. Harle had spent many years studying Eastern religions and mysticism that informs his thinking and artworks that are concerned with what he calls the '*techno*Metamorphosis' of humanity. He writes that all humans are now born into a technology saturated world that is affecting our future evolution despite our biologically hardwired predispositions. Harle explores *techno*Metamorphosis through digital visual art, academic research and writing, in the hope of arousing awareness in others of alternative scenarios inherent in this evolutionary process. The *techno*Metamorphosis includes not only the infiltration of devices like computers and mobile networks and their cultural impact, but even more so, the augmentation of existing biological humans with high-tech devices such as neural brain implants. It is somewhat paradoxical that Harle creates artworks that rely on digital/computer/technology to explore human-machine cyborgs while subtly asking the viewer, "Do you want this techno future for yourself and your children?"

I have addressed Harle's criticism in this new section of the postdigital edition of my book. My thesis that we are witnessing in postdigital art a paradigm shift from Hellenistic to Hebraic consciousness is enriched by exploring conceptual similarities between spiritual traditions of ancient cultures, spanning the Asian continent from Israel to India, Tibet, China, Korea and Japan. Have these similarities emerged from contacts between these different Asian civilizations millennia ago or are they expressions of the deep structure of human consciousness encoded in the DNA of all humanity? Nevertheless, the Hellenistic hegemony over the visual arts revived in Renaissance Europe has ended with the emergence of digital technologies in a networked world that links the new worlds of America and Australia with ancient worlds renewing themselves across Asia.

Synthetic reality at IBM Haifa

I took a break from writing this chapter to take a train from Tel Aviv to Haifa to participate in the Synthetic Reality Seminar 2010 at the IBM Research Center on Mount Carmel. I was the only artist at a conference on the science and applications of virtual, mixed, and augmented reality. For me, the presentations in non-art realms pointed to significant directions for postdigital art. The scientists and engineers presenting their research and development projects described how they were using digital technologies to rediscover ten fingers by creating active interplay between the realms of high tech and high touch. Digital codes composed to orchestrate haptic, kinesthetic, and social dialogues form feedback loops between cyberspace and real space.

The keynote speaker was Adrian David Cheok, director of the Mixed Reality Lab at the National University of Singapore and editor of *Advances in Human Computer Interaction*. His presentation, "Embodied Media and Mixed Reality for Social and Physical Interaction," described how researchers and students in his lab are expanding the interface between humans and computers that has not changed since the 1960s when the keyboard and mouse were invented and plugged into computers. They are studying ways to merge the virtual world into real-world space by allowing people to physically interact with computers in ways that feel natural and comfortable. Cheok's lab is not only creating digital systems with touch interfaces, but is developing biological systems that communicate social or ecological information and social network systems for intergenerational interaction through traditional cultural teachings.

Professor Cheok showed a wearable system aimed at promoting physical interaction in remote communication between parent and child and between people and their pets. In addition to parents talking to their children by mobile phone, this system enables them to hug one another through a hugging interface device and a wearable, hug reproducing pajama shirt connected through the Internet. The hugging device is a small, mobile doll with an embedded pressure sensing circuit that is able to accurately sense varying levels of the range of human force produced from natural touch. This device sends hug signals to a haptic jacket that simulates the feeling of being hugged to the wearer. It features air pressure actuation to reproduce hugging. Cheok showed a video of pet chickens, common in Asian countries, wearing haptic vests through which their owners could hug them through the Internet when away from home.

Cheok described how he uses an interactive living media to communicate social or ecological information in the form of an empathetic ambient media. In the fast paced modern world, people are generally too busy to monitor various significant social or human aspects of their lives, such as time spent with their family, their overall health, and the ecological state of their environment. This information quantified digitally is semantically coupled into living micro-organisms, *E. coli*. Using transformed DNA, the *E. coli* glow or dim according to the data. The core technical innovation of this system is the development of an information system based on a closed-loop control system through which digital input is able to control input fluids to the *E. coli*, and thereby control the output glow of the *E. coli* in real time.

The Mixed Reality Lab developed *Confucius Computer*, a cultural computing system based on the spiritual traditions of Asia that promote balance and harmony through intergenerational interaction. It is a postdigital system that brings to youth the traditions of their elders. This social network communication system uses new media to revive and model ancient Eastern philosophies and teachings, presenting them in new contexts. For example, it allows users to explore the traditional Chinese medicine concept of Yin-Yang through the interactive creation of new recipes for healthy meals. *Confucius Computer* also provides an artistic playground for interactive music-painting creations based on specially designed music filters and the ancient model of Cycles of Balance.

Tamar Weiss, professor of occupational therapy at the University of Haifa, spoke about "Technology Supported Collaboration Paradigms for Rehabilitation: From Physical to Virtual." In her research and practice, she has people in need of rehabilitation physically interact with displayed images, move and manipulate virtual objects, and perform other actions in a way that engenders a feeling of actual presence, and immerses their senses in the simulated environment. The user's actions in real space are mirrored by the actions of their virtual image as they receive visual, auditory, and sometimes haptic and olfactory feedback. A digital video camera converts the video signal of the user's movements into the same movements of the participant's live on-screen video image acting in real time. While navigating in a natural and intuitive manner within the virtual environment, the user can use all body parts to interact with virtual stimuli or interaction may be restricted to a specific body part. Weiss showed a video of a man needing therapy for the rehabilitation of his arms immersed as a virtual goalie using his arms in real space to deflect the balls coming at the video image of his body on a virtual soccer field. When he lifted his arm to stop a ball racing towards the goal, his virtual self lifted its arm and stopped the ball. In another version of this rehabilitation process, a woman sees herself in a pastoral 3D landscape with different colored balls coming down at her from the sky. She exercised her arms by making her virtual self reach up to touch the balls which instantly turned into birds that flew away.

Inspired by the Academy Award-winning film *Awakenings*, Yoram Baram, professor of computer science at the Technion-Israel Institute of Technology, created an augmented reality device to improve the distorted gait of movement disorder patients. In the film, a doctor played by Robin Williams notices that a woman with Parkinson's disease could walk on a floor of black and white tiles. Strangely, her walking ability stopped cold as soon as she reached the end of the checkerboard floor pattern. The tile pattern provided continuous, stabilizing visual information that safeguarded against stumbling and falling while walking. Baram developed a tiny device attached to a patient's eyeglasses that displays a virtual image of a tiled floor that responds to the patient's motions rather than just providing a constant visual display. "The image reacts to the patient's motions just like in real life," Baram explained at the IBM Haifa seminar. "For example, when the patient stands in place, the virtual floor does not move, but when he begins to walk, the floor starts moving beneath him. When he turns, the image of the floor also turns. Yet all the while the patient feels like he is walking on a steady floor." Baram was equipped to create this device by his experience designing a mechanism for NASA to navigate low-flying helicopters around obstacles such as trees, buildings, and electrical poles. The concept of the design, which Baram applied to this device, is that the optical images of objects help the observer stabilize himself in space. In the video that Baram showed at the IBM conference, patients with Parkinson's disease and stroke victims who staggered with a distorted gait began walking with a more normal gait using his device.

Rhizome, Facebook, LinkedIn

A postdigital way of identifying artists whose artworks exemplify postdigital art is to use web-enabled social networks. I posted a call for artists who consider their artwork to be postdigital on the Rhizome, LinkedIn and Facebook arts groups, and YASMIN (a Mediterranean Rim network of people exploring the intersections of art, science, technology and culture). I received nearly one hundred responses to my calls for participation. The largest number of responses came from members of Rhizome, a dynamic, interactive platform dedicated to the creation, presentation, preservation, and critique of emerging artistic practices that engage technology. Rhizome draws attention to artists, their work, their perspectives and the complex interrelationships between technology, art and culture. It was founded in 1996 by Brown University professor of media studies Mark Tribe as an e-mail list subscribed to by some of the first artists to work online. He chose the title "rhizome," a botanical term for a plant's horizontal, underground stem that sends out a network of shoots and roots from its nodes – a powerful metaphor for postdigital media in our networked world. Since 2003, Rhizome became affiliated with the New Museum of Contemporary Art in New York with thousands of subscribers worldwide. Mark Tribe[36] describes postdigital art as art that moves beyond creating pixelized pictures or lifelike digital animation to art that expresses how technology shapes ways of seeing and behaving.

Artists whose artworks are posted on the Rhizome website must label their art with tags chosen from a list provided by Rhizome. These tags can provide clues about the nature of emerging postdigital art forms: appropriation, bio art, blog, code, collaboration, collaborative, community data, data visualization, globalization, Google, identity, immersive, interactive, internet, narrative, net.art, network, open source, participatory, process, social media, social space, surveillance, text, virtual, webcam, wiki. A group of tags describes each artwork. Below are some exemplary postdigital artworks.

Surveillance/Interactive/Social Space: Ruth Ron created her site-specific *Screen-Wall* installation at the *Liquid Spaces* exhibition at the Israel Museum in Jerusalem to challenge the conventions of public and private spaces in a museum. The service and private parts of the museum – archives, offices, guard booth – that are traditionally closed to the public, become the subject of Ron's postdigital *Screen-Wall* reversing the relationship of watching and being watched. Instead of surveillance systems keeping an eye on the visitors, the visitors digitally peek at the concealed parts of the museum. Ron embedded small flat monitors in the gallery wall that displayed images of missing parts of the wall. A section of the wall with peeling paint was continued on the monitor next to it; half of a real double electrical outlet was completed by the virtual image of the other half. The visitor drawing close to the wall to examine the confused continuum between real and virtual images triggers the transformation of the images into live feeds of service locations throughout the museum streaming into the embedded monitors via the Internet. Ruth explains her concept: "The opaque solidity of physical architecture is challenged by multiple layers of the screen, the image of the wall and the transition to live video feed. The distance between remote spaces in the museum collapses, and digital and visual continuity is created. The network portal extends beyond the properties of the flat digital screen to become a reactive window to unexpected places." She quotes Marcos Novak: "When bricks become pixels, the tectonics of architecture becomes informational."

Identity/Participatory/Process: Sang Nam, educated as both biologist and artist, is a professor of interactive media at Elon University in North Carolina. He integrates digital and analog technologies in artworks that address the special aesthetic requirements of blind people. The "identity" tag is relevant because Sang Nam relates to these two populations as representing alternative identities rather than disabilities. He translates visual art designed for a seeing culture into the languages of a haptic and auditory culture. *Temperature Wave* is a metallic board that has many cells in which temperature can be electronically modified and coupled with changing sounds to offer blind people the experience of a dynamic two-dimensional artwork accessed through haptic and auditory senses. Digital technology controls the changing temperatures that can be felt as waves moving across the surface of Sang Nam's postdigital "painting" while an audio system patterns sounds to correspond to the movement of the heat.

Collaborative/Participatory/Process: Golan Levin, director of the Studio for Creative Inquiry and professor of electronic time-based art at Carnegie Mellon University, created in collaboration with eight other artists *Dialtones (A Telesymphony)*, a participatory concert performance with sounds produced through the carefully choreographed ringing of the audience's own mobile phones. Before the concert, participants register their mobile phone numbers at a series of web terminals where new ringtone melodies are transmitted to their phones and their seating assignments are generated. During the concert, the audience's phones are dialed up by live performers using custom software that permits as many as sixty phones to ring simultaneously. Because the exact location and tone of each participant's mobile phone is known in advance, the *Dialtones* concert is able to present a diverse range of unprecedented sonic phenomena and musically interesting structures, such as waves of polyphony which cascade across the audience.

Community Data/Collaboration/Data Visualization: New media artist Cynthia Beth Rubin coordinated the *Cultural Heritage Artists Project of the Orchard Street Shul* in New Haven, a project using social networking to build a community of creative visual thinkers and researchers investigating various facets of how we document, interpret, and retell cultural legacies. Local artists joined with artists from throughout the United States who came to New Haven to engage the old synagogue building a moment before its renovation and experience the community. They stepped out of the privacy of their studios, moving beyond their individual singular visions, and beyond solely listening to their inner artistic voices without regard for the audience. Through e-mail, an online networking site, and live encounters at the Shul, the participating artists shared digital photographs, videos, and sound recordings, exchanged ideas and traded stories, and translated original data collected into three-dimensional scans. In her contribution to this collaborative art project, Rubin collaborated with Seth Lamberton in integrating a three-dimensional digital model of the Shul with images of Eastern Europe from where the founders had come to New Haven and established their congregation nearly a century ago.

Data Visualization/Virtual/Process: Michael Shaw explores the potential of computer-aided design and manufacture (CAD/CAM) and the prototyping equipment in Loughborough University's mechanical engineering department to create sculptural forms that respond to light in unexpected ways. His biomorphic inflatables project ephemeral shadows and reflections, whereas his translucent resin forms appear to glow mysteriously from within. Shaw's creative use of rapid prototyping technologies translates the virtual digital realm into real-world matter. Shaw wrote to me, in response to my Rhizome invitation to postdigital artists, to say that he believes that art at its best exceeds or transcends the means of delivery. In the digital realm where delivery is usually confined to a rectangular screen, the transcending of medium is even more vital to avoid the allure of newness and the superficial luster of innovations in technology dominating – the wow factor. Interesting territory will emerge when digital work fully casts aside the shackles of the computer as a means of delivery. "Instigated in the computer, yes, but existent and manifested elsewhere."

Social Space/Identity/Participatory: Jacob Tonski activated his degrees in computer science and media arts to create *I'm Taller Than Most of the People I Know* at UCLA in 2008. He designed and built a computer-activated, self-adjusting platform that makes everyone the same height, probing ideas of equality and the origins of power. It adjusts the height of the floor so that two or three individuals stand at the same height, gazing eye to eye. Participants experience a subtle but profound and intimate shift in perception that questions diversity and differences as sources of richness in human relationships in relation to roots of division and power that invisibly shape the micropolitics of interaction between people. Tonki writes that height in this case serves as a metaphor for all of our personal differences, hoping to foreground their ever-present and overlooked impact by momentarily, and synthetically, removing them.

Wiki/Participation/Collaboration: Jared Nielson is developing a wikiart project *Chunks of Sensation*, a randomly constructed, non-narrative, cinematic experience created collaboratively by video artists worldwide. Modeled after the multiple authorship process of creating *Wikipedia*, this participatory art project attempts to establish the grammar of a new language that builds a bridge between cinema and television and the future of moving images. Chunks are short segments of video. When one chunk ends, another is immediately, randomly selected, ad infinitum, thus creating a "film" that plays forever. Viewers can enter and exit at any point and repeated viewings will be entirely different. This is a video equivalent to Nielson's 2008 musical collaborative project *The Song That Never Ends*, a randomly constructed experimental composition composed of hundreds of audio tracks contributed by musicians internationally. Each musician submits an audio track composed of one instrument that is added to a queue where it is randomly mixed with three more tracks by other musicians. When one track ends another is immediately selected and *The Song* plays forever.

Surveillance/Narrative/Internet/Google: Hasan Elahi, a professor at San Jose State University, is an interdisciplinary artist whose work examines issues of surveillance, simulated time, transport systems and borders. His whole life is an ongoing work about surveillance. He tells the chilling story of being stopped on returning from a trip from overseas to be interrogated by the FBI while being held in the INS detention center at the Detroit Airport. Over the course of questioning, it became clear that the reason he was being held was that he had a storage locker in Tampa, where he had been teaching. Scared by 9/11, the owners of the storage area reported that an Arab man had fled on 9/12 leaving explosives in his locker, although there were no explosives and he had not fled. In response to the FBI keeping after him for the following months and his providing his FBI agent with his routing before every trip he took, he decided why not tell not just the FBI, but everyone. So he hacked his cell phone into a tracking bracelet that he wore on his ankle, reporting his movements on a Google map. In addition to the mapping of his activities, he posts on the Internet at http://www.trackingtransience.net a photographic record of his daily life and a credit card record of all his purchases. He views his self-surveillance both as an art form and as his perpetual alibi for the next time the FBI questions him.

Autobiographic artworks like Hasan Elahi's are an increasingly common form of postdigital art. In his chapter "Lifelogging: A Concept of Sousveillance?"[37] in *(In)Visible: Learning to Act in the Metaverse*, Stefan Sonvilla-Weiss uses the term "sousveillance" for self-surveillance. Sometimes digital lifelogging is intertwined with other data visualizations as in Avi Rosen's *Digital Skin 2*. Rosen, a computer science researcher at the Technion-Israel Institute of Technology with a Ph.D. in art history from Tel Aviv University, traveled his digital sousveillance across the planet and into the endless expanses of the cosmos. His online video *Digital Skin 2* is a bricolage that combines cellphone photographs of himself and his activities with his endless virtual voyages in cyberspace on which he superimposes personal data on public databases (Google Earth and Sky). He transfers to public communications networks his body digital data strips mapping his real-world experiences merged with the earth and cosmos digital data strips produced by satellites and telescopes.

In fact, all the artists that I write about in this book engage in lifelogging on their websites and blogs telling the story of their creative lives. They also reveal their profiles on Rhizome, Facebook, LinkedIn, MySpace, or other social networks or through their avatars living in Second Life. Steiner and Yang in their book *Art Works: Autobiography*[38] survey public autobiographic narratives as artworks rather than just documentations of artistic careers. Artists' lifelogs often become autoethnographic studies of the artist's process of self-awareness in relation to the multiple cultures in which he lives, from the art world to his neighborhood. My *Artiststory* blog http://artiststory.blogspot.com, for example, continually updates my life as an artist in images and texts posted in the public domain. As a member of the Yahoo group "Autoethnography Listserv," I gain opportunities to join with people from many other fields of human endeavor to explore autoethnography as a qualitative method of research into one's own life in relation to cultural ecosystems. Rita Irwin[39] coined the term "a/r/tography" to describe this autoethnographic methodology in the life of the **A**rtist/**R**esearcher/**T**eacher. As is evident from the descriptions of artworks throughout this book, artists in the postdigital age also function as researchers and teachers. Artists-researchers-teachers inhabit and explore the borderlands between art, science, technology and education, integrating knowing, doing, and making through aesthetic experiences that flow between intellect, feeling, and practice to create and convey meaning. Irwin describes how a/r/tographers search for new ways to understand their art-making, research, and teaching through attention to memory, identity, reflection, meditation, storytelling, interpretation and representation. My chapter "From Awesome Immersion to Holistic Integration"[40] in *Educating Artists for the Future: Learning at the Intersections of Art, Science, Technology, and Culture* is an a/r/tographic journey from my childhood summers when I was set free among the sowbugs, salamanders, and swallows in the Catskill Mountains of upstate New York to my creating the curriculum for an art college in Israel in which students redefine art at the interdisciplinary interface where new technologies and scientific inquiry contribute towards shaping cultural values of a Jewish state in an era of globalization.

Tune out, turn off, unplug

I was excited about the artistic possibilities of digital technologies when the first computer plotter arrived at NYU's Courant Institute of Mathematical Sciences in 1965. I pleaded with the Institute's administrators to let me experiment with making art with their mammoth computer to no avail. "Giant super-expensive computers are for serious purposes, not for playing at making art," was their response. As a scientist in the process of metamorphosis into an artist, I was not unfamiliar with the world of computers. Indeed, I had written a paper "The Binary System and Computers"[41] describing how to build a simple digital computer that was published in the National Science Teachers Association journal *Science and Children* in 1964. In addition to being a doctoral student at NYU and studying painting at the Art Students League, I was working as science supervisor for a Long Island school system and test center coordinator for the American Association for the Advancement for Science (AAAS) curriculum project *Science: A Process Approach.*

When I spoke about my rejection to my doctoral advisor, Morris Shamos, head of the physics department and subsequently president of the National Academy of Sciences, he jumped with enthusiasm at my proposal to create art with computers. He phoned his colleague, the director of the Courant Institute, and explained to him the importance of humanizing digital technologies at a time when science fiction writers and TV show producers were frightening people into believing that computers would take over the world and enslave them. I was granted free reign of the computer center in response to Professor Shamos' question, "Is there a better way to put a human face on computers than to have them make art?" I programmed instructions for the computer to plot geometric pictures on rolls of paper. My cold, calculated computer-generated drawings seemed to invite a warm, sensuous, high touch response. The millennia-old technology of encaustic painting popular in ancient Greece and Egypt appeared to me to be a most appropriate high touch partner for high tech digital drawings. I made encaustic paints by melting beeswax with microcrystalline wax and damar resin crystals that I mixed with powdered pigments. The vibrant colors of my molten paints sensuously flowed over the plotter's hard-edged drawings hardening into a gem-like translucent surface. One of my early computer-generated encaustic paintings based on algorithms describing noise control was reproduced as the cover for the April 1966 issue of *International Science and Technology.*

The scary stories about humanity being enslaved by computers that science fiction writers wrote a half-century ago appear to be coming true in our present ubiquitous digital world that invades so many aspects of our lives. In his book *The Tyranny of E-mail: The Four-Thousand-Year Journey to Your Inbox*,[42] John Freeman describes how the computer and e-mail were sold to us as tools of liberation, but they have actually inhibited our ability to conduct our lives mindfully, with deliberation and consideration. By 2011, less than four decades after the first e-mail was sent, there will be more than three billion e-mail users. The flood of messages is ceaseless and follows us everywhere. AOL's 2008 E-mail Addiction Poll reported that 62% respond to e-mail on vacation, 67% check their e-mail in bed in their pajamas, and 60% read their e-mail sitting on the toilet. In addition, cellphone users in the USA alone in December 2008 wrote more than 110 billion text messages. Freeman ends his presentation of research evidence of the rapidly growing domination of our lives by digital technologies with the call "It's time for a break." Pulitzer Prize-winning author Geraldine Brooks writes in her review of *The Tyranny of E-mail*, "Freeman's impeccably researched, eloquently argued book reveals the many ways this so-called boon to communication and productivity has become a distracting, privacy-sapping, alienating, addicting time-suck. He has convinced me that the new mantra for our times ought to be Tune out, Turn off, Unplug."

A powerful antidote to being enslaved was articulated nearly 3,500 years ago as the Israelites gathered at the foot of Mount Sinai shortly after their exodus from centuries of slavery. The Ten Commandments that my ancestors received there proposed a Sabbath day as a potent way to take a break. As slaves under Egyptian oppression, they were forced to work incessantly with no breaks. They were not free to choose what to do with time that was not their own, time that was owned by their slave masters. *Shabbat*, a day of rest every seventh day, was a revolutionary concept in the ancient world that became an integral part of Hebraic consciousness until today. Observant Jews continue to keep the Sabbath by tuning out, turning off, and unplugging. From sunset on Friday until stars dot the sky on Saturday night, I disconnect from the digital world. I turn my computer off. I do not send e-mails, do not tweet on Twitter, do not write on Facebook walls, do not google, do not blog, link out of LinkedIn, tune out of YouTube, and leave the virtual world of Second Life for my real-world first life. For 25 hours each week, there are no ringing phones, no radio or TV, no traveling on the information highway and no traveling on the asphalt highway. Pollution from information overload and carbon emissions are stopped cold on the day of rest. Shabbat is Ecology Day, a day we leave the world the way we got it, a joyous day set aside to take pleasure in divine creation while enjoying each other. Shabbat is also a Non-Art Day on which I stop making all art – postdigital, digital and pre-digital. All activities inappropriate on Shabbat are derived from the 39 craft categories that went into making the Tabernacle.

Shabbat was given to the Israelites at Sinai as a gift for all humanity, a gift particularly valuable to everyone in our fast-paced postdigital world. In *The Sabbath: Its Meaning to Modern Man*, Rabbi Abraham Joshua Heschel teaches that observing Shabbat invites us to become attuned to holiness in time rather than living in the tyranny of space, both real and virtual. "It is a day in which we are called upon to share what is eternal in time, to turn from the results of creation to the mystery of creation; from the world of creation to the creation of the world."[43]

Observance of Shabbat is in tune with *Wikipedia*'s definition of postdigital as "an attitude that is more concerned with being human, than with being digital." It is enjoying family and friends and making love. My wife and I celebrate Shabbat with a Thanksgiving dinner, not once a year in November, but every week. We enjoy our growing family of children, grandchildren, and great-grandchildren singing together around our table covered with a tablecloth embroidered by my mother and a *h̲alah* cover embroidered by my wife's mother illuminated by the warm glow of the Shabbat candles. The highest joy is the traditional Shabbat observance of making love. Rabbi Pincas Peli in *The Jewish Sabbath: A Renewed Encounter* explains the significance of Shabbat as an ideal time for intimacy between husband and wife:

> To have sexual intercourse is, in biblical language, to *know* someone. It is the act that combines mutual understanding and physical attraction: desire and relationship culminate in an act of union with another person that touches both partners in body and soul. Ideally, it is the confirmation in action of a total acceptance and affirmation of the uniqueness and spirituality of Shabbat. At the same time, the physical well-being, lack of pressure, and leisure of Shabbat is the perfect setting for attaining such a high degree of intimacy and openness to the other. In turn, the union enriches and deepens the individual's sense of well-being and value. Thus it strengthens the process of becoming more human on this day.[44]

The fourth of the Ten Commandments enjoins us to observe the Sabbath day to make it holy. We can work for six days, but on the seventh day we are not to do anything that constitutes work. We must remember that because God freed us from slavery in Egypt, we are commanded to keep the Sabbath (*Deuteronomy 5:12–15*). All humanity is invited to observe Shabbat as a powerful way to free us from being enslaved to our own creations and from our addictions to new technologies. It is a divine design to help make us all more human. And on the eighth day, we can return with renewed energies to being the partners of God in continuing creation. We enjoy being immersed in the amazing technological wonders of our era knowing that we are free to tune out, turn off, and unplug on the next Shabbat.

Notes

1. Roy Ascott, "Pixels and Particles: The Path to Syncretism," in Mel Alexenberg (ed.), *Educating Artists for the Future: Learning at the Intersections of Art, Science, Technology, and Culture* (Bristol and Chicago: Intellect Books/University of Chicago Press, 2008), pp. 47–60.
2. Stephen Wilson, "Beyond the Digital: Preparing Artists to Work at the Frontiers of Technoculture," in Mel Alexenberg (ed.), *Educating Artists for the Future: Learning at the Intersections of Art, Science, Technology, and Culture* (Bristol and Chicago: Intellect Books/University of Chicago Press, 2008), pp. 29–46.
3. Nicholas Negroponte, "Beyond Digital," *Wired*, December 1998, Issue 6, p. 12.
4. Postdigital, *Wikipedia*, "Postdigital", http://en.wikipedia.org/wiki/Postdigital. Accessed October 10, 2010.
5. Menachem M. Schneerson, *Toward Meaningful Life: The Wisdom of the Rebbe*, adapted by Simon Jacobson (New York: William Morrow, 1995), p. 191.
6. Allan Kaprow, *Essays on the Blurring of Art and Life*, Jeff Kelley (ed.) (Berkeley: University of California Press, 1993), p. 223.
7. Craig Roland, *The Art Teacher's Guide to the Internet: Ideas, Tools, and Resources for Teaching Art and Design in a Post-Digital Age*, http://artjunction.org/blog/?p=920. Accessed Ocotber 7, 2010.
8. *Wiktionary*, "Postdigital", http://en.wiktionary.org/wiki/postdigital. Accessed October 7, 2010.
9. Robert Pepperell and Michael Punt, *The Postdigital Membrane: Imagination, Technology and Desire* (Bristol: Intellect Books, 2003).
10. Daniel H. Pink, *A Whole New Mind: Why Right-brainers will Rule the Future* (New York: Riverhead Books, 2006), pp. 51–52.
11. Thomas Friedman, *The World is Flat: A Brief History of the Twenty-First Century* (New York: Farrar, Straus and Giroux, 2006).
12. Mel Alexenberg (ed.), *Educating Artists for the Future: Learning at the Intersections of Art, Science, Technology, and Culture* (Bristol and Chicago: Intellect Books/University of Chicago Press, 2008).
13. Stephen Wilson, *Information Arts: Intersections of Art, Science, and Technology* (Cambridge, MA: MIT Press, 2002).
14. Roy Ascott, *Telematic Embrace: Visionary Theories of Art, Technology, and Consciousness* (Edward A. Shanken ed.) (Berkeley: University of California Press, 2003).
15. Stefan Sonvilla-Weiss, *(In)Visible: Learning to Act in the Metaverse* (New York: SpringerWien, 2008).
16. Eduardo Kac, *Signs of Life: Bio Art and Beyond* (Cambridge, MA: MIT Press, 2007).
17. Yacov Sharir, "Learning through the Re-embodiment of the Digital Self," in Mel Alexenberg (ed.), *Educating Artists for the Future: Learning at the Intersections of Art, Science, Technology, and Culture* (Bristol and Chicago: Intellect Books/University of Chicago Press, 2008), pp. 217–28.

18. Eduardo Kac, "Life Transformation - Art Mutation," in Mel Alexenberg (ed.), *Educating Artists for the Future: Learning at the Intersections of Art, Science, Technology, and Culture* (Bristol and Chicago: Intellect Books/University of Chicago Press, 2008), p. 212.
19. Christiane Paul, *Eduardo Kac: Lagoglyphs, Biotopes and Transgenic Works* (Rio de Janeiro: Oi Futuro, 2010), http://www.ekac.org/oi_kac.show_overview.html. Accessed October 7, 2010.
20. Ibid.
21. Svetlana Alpers, *Rembrandt's Enterprise: The Studio and the Market* (Chicago: University of Chicago Press, 1988), p. 105, 110.
22. Thomas L. Friedman, *The Lexus and the Olive Tree* (New York, Farrar, Straus and Giroux, 2000), p. 41.
23. Ibid., p. 31, 41.
24. James Clifford, *Routes: Travel and Translation in the Late Twentieth Century* (Cambridge, MA: Harvard University Press, 1997), pp. 247–48, and Nicholas Mirzoeff, "Multiple Viewpoints: Diaspora and Visual Culture," in Nicholas Mirzoeff (ed.), *The Visual Culture Reader* (London and New York: Routledge, 2002), pp. 206–08.
25. Thorleif Boman, *Hebrew Thought Compared with Greek* (New York: Norton, 1960), p. 27.
26. Dan Senor and Saul Singer, *Start-up Nation: The Story of Israel's Economic Miracle* (New York and Boston: Twelve, A Council on Foreign Relations Book, 2009).
27. Ibid., p. 51.
28. George Gilder, *The Israel Test* (Minneapolis: Richard Vigilante Books, 2009).
29. Avigdor Shachan, *In the Footsteps of the Ten Lost Tribes* (Jerusalem: Devorah, 2003).
30. Mel Alexenberg, "Ancient Schema and Technoetic Creativity," *Technoetic Arts: A Journal of Speculative Research*, 4: 1, 2006, pp. 3–14.
31. Wengao Huang, "New Media Art as Embodiment of Tao," in Mel Alexenberg (ed.), *Educating Artists for the Future: Learning at the Intersections of Art, Science, Technology, and Culture* (Bristol and Chicago: Intellect Books/University of Chicago Press, 2008), pp. 155–68.
32. Rodger Kamenetz, *The Jew in the Lotus* (San Francisco: HarperSanFrancisco, 1994), p. 200.
33. Diane Gromala and Jinsil Seo, "Touching Light: Post-Traditional Immersion in Interactive Artistic Environments," in Mel Alexenberg (ed.), *Educating Artists for the Future: Learning at the Intersections of Art, Science, Technology, and Culture* (Bristol and Chicago: Intellect Books/University of Chicago Press, 2008), pp. 175–90.
34. Vinod Vidwans, "Expressing with Grey Cells: Indian Perspectives on New Media Arts," in Mel Alexenberg (ed.), *Educating Artists for the Future: Learning at the Intersections of Art, Science, Technology, and Culture* (Bristol and Chicago: Intellect Books/University of Chicago Press, 2008), pp. 141–54.
35. Rob Harle, "The Future of Digital Art: From Hellenistic to Hebraic Consciousness," *Leonardo Reviews*, December 2006, http://www.leonardo.info/reviews/dec2006/future_harle.html. Accessed October 7, 2010).
36. Mark Tribe as told to Alex Gartenfeld, "For the Love of Art: Mark Tribe on Post-Digital," *PaperMag*, http://www.papermag.com/?section=article&parid=2900. Accessed October 7, 2010).
37. Stefan Sonvilla-Weiss, "Lifelogging: A Concept of Sousveillance?" in *(In)Visible: Learning to Act in the Metaverse* (New York: SpringerWien, 2008), pp. 73–101.
38. Barbara Steiner and Jun Yang, *Art Works: Autobiography* (New York: Thames & Hudson, 2004).
39. Rita Irwin and Alex de Cosson (eds.), *A/r/tography: Rendering Self through Arts-based Living* (Vancouver, BC: Pacific Educational Press, 2004).

40. Mel Alexenberg, "From Awesome Immersion to Holistic Integration," in Mel Alexenberg (ed.), *Educating Artists for the Future: Learning at the Intersections of Art, Science, Technology, and Culture* (Bristol and Chicago: Intellect Books/University of Chicago Press, 2008), pp. 305–35.
41. Mel Alexenberg and R. B. Blaney, "The Binary System and Computers," *Science and Children*, 2: 3, 1964.
42. John Freeman, *The Tyranny of E-mail: The Four-Thousand-Year Journey to Your Inbox* (New York: Simon & Schuster, 2009).
43. Abraham Joshua Heschel, *The Sabbath: Its Meaning to Modern Man* (New York: The Noonday Press/Farrar, Straus and Giroux, 1951), p. 10.
44. Pincas H. Peli, *The Jewish Sabbath: A Renewed Encounter* (New York: Schocken Books, 1988), p. 86.

Chapter 2

Semiotic Perspectives: *Redefining Art in a Postdigital Age*

The eye, so prized an aesthetic organ when it was felt that the difference between art and non-art was visible, was philosophically of no use whatever when the differences proved instead to be invisible. Pop artists were joined by the Minimalists in showing that there is no special way a work of art has to look. It can look like a Brillo box if you are a Pop artist, or like a panel of plywood if you are a Minimalist[...] What makes the difference between art and non-art is not visual but conceptual.[1]

Arthur Danto

there is an easy answer to the question why "new tendencies in art" are always rejected (and, at times, with what a fuss!): because the eye, accustomed to earlier "recipes," is not immediately capable of perceiving the newly discovered "recipes." it has to have time to get used to them. Something I might describe as "conservatism of the eye."[2]

Wassily Kandinsky

We are experiencing "The End of Art" as the visual perception of surface gives way to the conceptual grasp of inner significance. In *Beyond the Brillo Box: The Visual Arts in Post-Historical Perspective*, Columbia University philosophy professor Arthur Danto discusses how Andy Warhol's 1964 exhibition at the Stable Gallery in New York marks the end of art. In the art gallery, Warhol stacked boxes on which he had screen-printed the Brillo logo. They looked identical to the cartons of Brillo soap pads that we see in supermarket aisles. We could no longer *see* the difference between *Brillo Boxes* (the work of art) and Brillo boxes (the mere real things). What Warhol taught was that there is no way of telling the difference by merely looking. The history of Western art as a progressive historical narrative of one art style superseding a previous style came to an end.

I believe that what we are witnessing is not the end of art, but the end of art derived from a Hellenistic structure of consciousness. The contemporary redefinition of art is emerging from a Hebraic biblical consciousness as expressed through the oral Torah. Danto's radical new proposal that concept and context rather than visual appearance gives meaning to images and objects was seriously discussed centuries ago by rabbis dealing with idolatry and Greek aesthetics. In the Talmudic tractate *Avodah Zarah* (Strange Worship), rabbis discuss whether found fragments of an image such as the hand or foot of a statue that was worshipped are prohibited or permitted. If the idol fell down and broke, Rabbi Shimon ben Lakish reasoned, then the hand or foot are permitted because the owner of the idol annuls it by saying, "If it could not save itself, so how could it save me?" Samuel explained that if they were mounted on a pedestal they were still valued as idols. Therefore, the exact same hand or foot would be prohibited.[3]

The Greek Proclos, son of a philosopher, put a question to Rabbi Gamliel who was bathing in a pool in front a large statute of Aphrodite. "If your *Torah* forbids idolatry, why are you bathing in the Bath of Aphrodite?" The rabbi answered, "I did not come into her domain, she came into mine." If the statue of Aphrodite was erected and then a pool was made to honor her, it would be forbidden for a Jew to bathe there. However, if the pool was made first and the statue was placed there as an adornment, then it is permitted.[4] Concept and context determine meaning in the case of the idol fragments and the statute of Aphrodite, like Brillo boxes in an art gallery rather than in a supermarket and a panel of plywood hanging in a museum rather than stacked in a lumberyard. The visual sense alone cannot discern between art and non-art today or between idol and mere decoration yesterday.

From representational to presentational art

A decade after Warhol's Stable Gallery show, I found that my students at Columbia University had a difficult time making sense out of the multiple new directions in art that they found happening throughout New York. The groundbreaking art forms they were experiencing in galleries, alternative spaces, performance art venues, on the street and in the media, demanded a redefinition of art. In response to this demand, I developed a scheme for classifying these innovative forms of art in relation to older forms.[5] I based my taxonomy on the work of Charles Peirce, the late 19th-century Harvard philosophy professor who is considered the father of semiotics – the theory of signs and how they create significance. He identified three types of representational signs: icon, symbol and index.[6] ***Re***presentation creates significance through resemblance, consensus, and documentation of after-the-fact signs of what was.

Since I found Peirce's three semiotic classes insufficient when applied to modern and postmodern art forms, I created three classes of presentational art that located art in the present and future in addition to the categories of representational art that located art in the past. These classes of presentational art are identic, prioric and dialogic. Identic art gains meaning by presenting what is. Prioric art presents what can be. Dialogic art gains meaning through dialogue, collaboration, and interaction in a dynamic process of becoming. A work of art can gain meaning through a single mode or any combination of modes of signification. Understanding relationships between the six semiotic classes of art can give us conceptual tools for tracing the evolution of the shift from Hellenistic to Hebraic definitions of art, from representation to presentation, and from icon to dialogue. Applying the three classes of presentational art can help us understand how the impact of digital technologies on postmodern art creates aesthetic significance. Digital art forms and high touch reactions to high tech experience also refresh our understanding of Peirce's classes of representation.[7]

Iconic art: resemblance

Iconic art represents the external appearance of things. It gains meaning by looking like something that we see in the real world. Redon's painting of a vase of flowers, Michelangelo's Adam reaching out to touch the hand of God, Picasso's *Three Musicians*, and a road map are all icons with different levels of iconicity. Iconic art can range from the Greek art ideal of creating an illusion so great that birds are fooled into eating painted grapes, from *trompe d'oeil* still lifes, and from photorealist cityscapes, to abstracted objects and schematic drawings. Computer users know the word "icon" as the images of a blank sheet of paper with its corner folded down, the floppy disc, the file folder, the printer, and the scissors icons on the toolbar of computer screens.

Computer simulations can render icons so realistically that they create a virtual world visually indistinguishable from the real world. When I was at the 1984 SIGGRAPH computer imaging and animation convention in Minneapolis, the film *The Last Starfighter* had just been released. It starts with high touch images of a backwater California trailer park, a warm, friendly, black, elderly gentleman and sleepy basset hound. Even the neon sign announcing the name of the trailer park had a high touch glow in contrast to the high tech dazzle of the sleek, shiny, metallic spaceships that we see later. New digital technologies make the old electric technology quaint. The park's teenagers' major source of entertainment was a videogame, *The Last Starfighter.* One of the kid's making a perfect score in the game was the big exciting event in the laidback trailer park. The teenagers had no idea that the videogame was placed in different places on planet Earth to identify naturally gifted starfighters for an intergalactic war. The trailer park videogame hero was recruited for the good guys in the war. When he arrived at the distant planet under siege, all seemed lost. There was only one starship remaining to save the defenders of good against the evil armada. The Earthling teenager protests being put into the cockpit of the high tech starship, claiming that he only knows how to drive a pickup truck. He is assured that flying the real starship is identical to playing the videogame simulation at which he proved to be the very best. With amazing skill learned on the flight simulator in the California trailer park, he destroys the enemy armada with his single starship and saves the entire galaxy.

Beyond painted grapes that pigeons come to eat, digital technologies have upgraded iconic representations to simulations in which the real and the imitation are indistinguishable. Indeed, computer graphics images that imitate photographs have become more realistic than the photographs themselves. In the film *Jurassic Park*, the computer animators made the dinosaurs so perfect that they were too real. Lev Manovich explains what needed to be done to icons with such a high-level of iconicity that they become unbelievable.

> Typical images produced with 3D computer graphics still appear unnaturally clean, sharp, and geometric looking. Their limitations especially stand out when juxtaposed with a normal photograph. Thus one of the landmark achievements of *Jurassic Park* was the seamless integration of film footage of real scenes with computer-simulated objects. To

> achieve this integration, computer-generated images had to be degraded; their perfection had to be diluted to match the imperfection of film's graininess.[8]

Computer-simulation of high-level iconic images set them in motion in movies and computer games in contrast with iconic representation in painting and sculpture which rendered movement static. Arresting motion in Western art is derived from iconic representation valued in the ancient Greek world, which suspended motion and made life still. Motion was activated, however, in Hebraic paintings 1,800 years ago in the Dura Europos synagogue on the banks of the Euphrates. These fresco-paintings sequenced iconic images in a continuous narrative like comic strips or frames of computer animation. Art historian Ernest Namenyi explains:

> The most striking feature of the Dura Europos synagogue's murals is that the representation of persons and beauty of the human form remain of obviously secondary importance to the artist, who seems not to have been at all concerned with three-dimensional representation and to have felt little interest for the tricks of perspective that he had learned from Greek artists. The general setting of action, its architecture and landscape, are his main preoccupation, and to them he devotes his full attention. The element of motion in the spectacle that he thus provides is developed to such an extreme that it distracts our attention from each human figure in turn.[9]

It is appropriate at this point to examine the common misunderstanding about the Hebraic aversion to iconic representation and alleged hostility to the visual arts emanating from the second of the so-called Ten Commandments. Elucidation of the semiotic classification scheme for contemporary art will make it clear how a Hebraic aesthetics is best expressed through dialogic art forms in contrast with Hellenistic aesthetic values that honor the iconic. Hebraic aesthetic values are primarily about representing active processes in which time informs space, and presenting opportunities for dynamic dialogue, expansive integral thought, and interactive experience in open-ended systems.

We can best understand the second commandment in the context of the others. First let use make clear the inaccurate translation of the Hebrew *dibrot* which does not mean "commandments" at all, but rather "utterances." The Bible describes ten Divine Utterances. *Dibrot* is from the verb *l'daber*, "to speak," and is related to the noun *davar*, "utterance, thing, idea, or act." Let us begin by translating the first of the *dibrot* from the original Hebrew (*Exodus 20:1–14*, repeated in *Deuteronomy 5:6–18*). "God spoke (*daber*) all these things (*d'vrarim*), saying: 'I am Was-Is-Will Be, your God, who has taken you out of the Land of Narrow Straits (*mitzrayim*, the Hebrew word for Egypt literally means "from narrows straits"), out of the house of slavery.'" The biblical divine name *YHVH* is a verb, not a noun. It combines the words for "was", "is", and "will be." God does not identify Himself here as the Creator of the world of space, but rather as a process acting in time to free us from enslavement to narrow-minded thinking.

"Do not have any other gods before Me. You shall not make yourself any carved statue or picture of anything in the heaven above, on the earth below, or in the water beneath the land." This second utterance, not exclusively addressed to artists, asks us not to transform dynamic process into static form. God is a verb that cannot become a noun. God is no thing and can only be represented as *no*thing. More than being commanded not to make an iconic representation of God, we are charged with recognizing the impossibility of depicting an invisible God. We can only become aware of divine light flowing in our world of everyday life. We are enjoined not to freeze the processes of creation and historical processes as fixed images, not to lower ourselves into patterns of narrow-minded thought, not to carve in stone that which is in flux.

"Do not take the name of Was-Is-Will Be, your God in vain" is the third utterance. It enjoins us not to limit the Infinite by assigning it a name. In everyday Hebrew conversation and in contemporary biblical translations, God is called *Hashem*, literally "The Name." It would probably be better, however, to call God *Bili Shem*, "Nameless." All the many divine names in Hebrew are not considered God's names, but rather terms for facets of the emanation of divine light into the world of time and space.

The fourth utterance asks us to keep the Sabbath by abstaining from doing creative work. Although we are enjoined from arresting the creative flow by freezing it in space, we do stop the flow in the realm of time. We stop our own creative process every seventh day to enjoy and honor divine creation. We stop our artistic activity to honor the Divine Artist on one day each week. We replenish our souls on the Sabbath so that on the eighth day, we are opened to reassuming the role of partners with the Creator in *tikun olam*, actively making the world a better place for all humanity. From when the sun sets on Friday evening to the time stars dot the sky on Saturday night, we celebrate Ecology Day when we honor nature and leave it undisturbed. We acknowledge that our mastery over our world is a divine gift.

Three additional biblical passages can help us understand the Hebraic aversion to static, iconic, closed systems and its appreciation for dynamic, dialogic, open systems. When Moses sent leaders of all the Israelite tribes from the desert to explore the land that God has promised them, they "headed up and explored the land from the wilderness of Tzin to Re<u>h</u>ov" (*Numbers 13:21). Re<u>h</u>ov* intimates "wide expanses." Having left slavery in the narrow straits, *Mitzrayim* (Egypt), they were headed toward the freedom of wide expanses, *Re<u>h</u>ov.* When the twelve scouts returned to report, ten had retained their narrowness of thought, their vision arrested by their slave mentality. They could not envision beginning a new life in freedom in their own land. They sought a new leader who would bring them back to *Mitzrayim* despite the objections of Moses' loyal disciple, Joshua.

Only Calev could open his eyes to freely see a land flowing with milk and honey. The Bible tells us that only Calev had a "different spirit" (*Numbers 14:24*). The ten narrow-minded scouts continued to wander in the desert for four decades and died there. Their children, born in the open expanses of the desert, followed Calev and Joshua into the land of Israel. Before entering the land, Joshua sent two scouts ahead saying, "Go, observe the land and Jericho. So they went and arrived at the house of a woman innkeeper whose name was Ra<u>h</u>av" (*Joshua 1:1*). *Ra<u>h</u>av* means wide opened. In summary, we can read *Psalm 118:5*, "From the narrow straits (*metzar*) I called upon God; God answered me with expansiveness (*mer<u>h</u>av*)." More than three millennia later, only the descendents of Calev of the tribe of Judah, preserved by their "different spirit," could enter the Promised Land in freedom to create the modern State of Israel.

Two of Calev's descendents, Komar and Melamid, whose spirit of freedom defied Soviet enslavement by creating Socialist Realist paintings that subverted the totalitarian regime that gave birth to that painting style. Socialist Realism, iconic art of the highest order, was idolized by Stalin and used to idolize him. In their 1983 painting, *The Origins of Socialist Realism*, Komar and Melamid painted Stalin as a modern-day Pharaoh royally seated before marble columns, red drapery, and a burning torch while being fondled by a half-naked nymph. Iconic art expresses Jewish sensibilities when it subverts idolatry.

When Komar and Melamid emigrated from Russia to settle in New York, they collaborated in painting the ultimate work of iconic art through their project *Painting by Numbers: A Scientific Guide to Art.*[10] They hired a polling firm to find out what Americans most wanted in art through telephone interviews of a statistically significant population sample. Americans wanted blue landscapes with historical figures and wild animals. The poll results where accurate within a margin of error of +/- 3.2 percent, at a 95 percent level of confidence. Based on the poll results, Komar and Melamid painted "America's Most Wanted" painting.

Subverting idolatry with a twist of irony has been the mission of the Jews from their very beginning. As a prelude to the biblical story of Abraham beginning his journey away from his birthplace and his father's world of idolatry, the *Midrash*[11] tells that Abraham was minding his father's idol shop when he took a stick and smashed the merchandise to bits. He left only the largest idol untouched placing the stick in its hand. When his father returned, his shock at seeing the scene of devastation grew into fury as he demanded an explanation from his son. Abraham explained how the largest idol had broken all the other idols. He could have smashed all the idols without saving one on which to place the blame. An idol smashing idols gives us clues for creating art to debunk Art.

In *Idolizing Pictures: Idolatry, Iconoclasm and Jewish Art,* Anthony Julius proposes that the primary role of the Jewish artist is to subvert idolization of totalitarian leaders and political systems as well as art itself. Jewish art aims to undermine undue reverence for art. The most common form of modern idolatry is when the work of art or the State become idols, alienated from their makers and given a false sovereignty. In its postmodern form, Jewish artists attempt to use art to knock art off its pedestal by displaying a creative skepticism not just towards art's subjects but also towards its purposes. "By creative skepticism I mean something like an art-making iconoclasm, that is, an art which turns against Art."[12]

Symbolic art: consensus

After icon, the second semiotic class is symbol. Symbolic art represents things or ideas through signs that are assigned meaning maintained by convention, by the agreement of community. Unlike an icon that bears a likeness to what it signifies, a symbol bears no direct or necessary connection to what it signifies. A red traffic light, for example, signifies a command to stop while a green light signifies go. These are assigned meanings agreed upon by community consensus. Had the opposite assignment been made, green would signify stop.

I have shown a slide of Larry River's painting, *Last Civil War Veteran*, when I lectured in Israel, Holland and Japan. No one could identify the subject of the painting that shows the Confederate and Union flags behind an abstracted figure in a bed. They all recognized the Union flag as the flag of the United States of America, but none could recognize the flag of the Confederate states. On the other hand, when I showed this same slide in the USA, everyone could identify the subject of the painting. A flag has symbolic significance when the country it signifies is known in advance.

The supreme symbol is the word. Only if you know the language, can you read and understand the word. If an artist makes a painting of a man walking on the beach that clearly looks like a man walking on the beach, it is iconic art. If an artist paints the words EEN MAN WANDELT OP HET STRAND in blue letters on an ochre background on a large canvas, it is symbolic art. However, the words in the painting are semantically meaningless unless you understand Dutch. They mean: "Man Walking on the Beach."

Paper money is the supreme example of symbolic value. It has no intrinsic value. It represents value by consensus of a group of people who agree that it can be exchanged for commodities. The Dutch in Rembrandt's time made art, like money, a symbol of economic value. Portable paintings on canvas could be traded by merchants in a free market economy instead of immovable painted walls commissioned by clergy or aristocracy for church or palace. Svetlana Alpers documents Rembrandt's role in transforming artworks into commodities in *Rembrandt's Enterprise: The Studio and the Market.* Rembrandt chose the exchange system of the marketplace to free him from having to seek honor in the courts of patrons.

> He pursued honor not in the sense of honors that others could confer, but in the sense of what art itself could confer, the value that his art itself brought into being, and this was registered in the money values of the market [...] His paintings shared with money – a piece of metal or paper, marked with certain symbols – a quality which economists refer to as abstract: though nothing in itself, it is accepted as the representation of value.[13]

Rembrandt would have been amused at the Dutch committee of Rembrandt experts, more than three centuries later, ruling that the famous Rembrandt painting, *Polish Rider*, owned by the Frick Collection was not a real Rembrandt after all. Because of this de-attribution,

The New York Times reported that its value fell to one-sixth of its initial value in one day. The painting, although unchanged materially, lost its symbolic value. At the time, I was head of the fine arts department at Pratt Institute and was teaching a course that I called, "Fine Arts with Computers." I scanned a reproduction of *Polish Rider* and digitally squashed it to one-sixth of its original width into a Giacometti-like skinny, elongated horse and rider.

Andy Warhol went one step further than Rembrandt in his 1962 money pictures in which he silkscreened images of dollar bills on canvas. He enjoyed selling his artistic symbols of US Treasury symbols for more money than the face value of the bills.[14] In the digital age, symbolic money is further abstracted from atoms to bits. Atoms of gold with some intrinsic value have changed to atoms of printed paper with only symbolic value, to bits of electronic information that can flow around the planet as a Visa card is swiped through the digital reader of the magnetic strip. We are witnessing the trend Lucy Lippard calls "the dematerialization of the art object,"[15] paralleling the dematerialization of money. With the click of a mouse, we can make our bid for a Rembrandt etching at an Internet art auction. Bits buy atoms.

Beyond buying atoms with bits, residents of virtual worlds in a postdigital age are using symbolic virtual money to buy virtual things rather than real things. In Second Life, avatars can use virtual Linden dollars to buy virtual Rembrandt paintings to hang on the virtual walls of their virtual homes on virtual tropical islands. Fantasy-world Linden dollars can be exchanged for real-world US dollars deposited directly into PayPal accounts. In 2009 as the world economy wilted, Second Life's economy blossomed as transactions between this virtual world's residents more than doubled in one year to the equivalent of $567 million in actual US currency. Second Life's shopping webpage entices its residents: "Change your shoes or outfit as often as you change your mind. Maybe today you'll be a robot or a tiny rabbit, or role-play a fantasy character with your friends…" Since Second Life does not have to deal with pesky issues like space, time or scarcity of resources, you can buy just about anything you can imagine. There is a huge, thriving fashion industry creating the hottest styles for you to wear. And personal style doesn't stop with clothes – you can also buy a new body, new hairstyles, and new animations for your avatar. In Second Life, you can transform yourself into an avatar of Rembrandt himself.

The real Rembrandt sold his painting, *Aristotle with a Bust of Homer*, to the Sicilian nobleman Don Antonio Ruffo for 500 guilders, eight times what he would have paid a first-rate Italian painter, and as much as ten times what Rembrandt could have gotten from a Dutch collector.[16] It is now worth millions, one of the most valuable symbols of Western art in the world. It hangs in the Metropolitan Museum of Art contained in an elaborate gold frame with a velvet-covered rope draped in front of it to keep the public back. A guard is posted to insure that we only look from afar and not be tempted to run our fingers over the inviting, sensuous brush strokes of the master. The great Greek philosopher contemplating a Greek bard's bust can be perceived by the visual sense alone, isolated from all other senses. This visual attention to surface is the highly prized Hellenistic value that had informed Western aesthetics. Kant's view is that "entirely disinterested satisfaction is called beautiful."[17] Bullogh proposes that aesthetic experience results from "psychical distance," visual contemplation of art detached from all other concerns.[18] Psychological distance is expressed through physical distancing. Paying homage to highly valued art objects held in distant awe is alien to Jewish sensibilities, coming too close to idolatry for comfort.

To document this visual distancing, my art students at Columbia University made a video of what appears to be a ritual of idolatry at the Metropolitan Museum of Art, an edifice imitating the architecture of ancient Greece. People ascend the broad steps before this neo-Grecian temple to the world's great art. Inside, they wait in line to pay a tithe, an entitlement to enter the inner sanctuary. They ascend an internal flight of marble stairs, stop, look at the honored painting, walk closer, bow before the painting (to read the label next to it), back away from the painting, and stand in silent adoration while viewing it from a distance. This ritual of homage is then repeated at the next painting.

Moving beyond creating art for individuals to experience from a distance in awesome silence, artists in our postdigital age are creating "new forms emphasizing our essential interconnectedness rather than our separateness, forms evoking the feeling of belonging to a larger whole rather than expressing the isolated, alienated self."[19] Like new forms of performance art and participatory art, Jews relate to their most valued possession, the Torah scroll, with intimate contact. Three days a week, we undress the scroll removing the velvet mantle to reveal columns of handwritten text that are chanted aloud in public. When the annual cycle of reading the Torah is completed, we dance hugging the scroll, holding it close to our bodies while singing. We kiss it as we pass it on from dancer to dancer. Freeing a Rembrandt painting from its frame, rolling it up, hugging and kissing it while dancing with it in song down Fifth Avenue would land you on a psychiatrist's couch or in a jail cell.

As noted above, the symbolic significance of the spiral scroll form is so strong that if we do not have a Torah scroll in the synagogue, we do not read the biblical portion at all. We do not read the exact same words from a Bible printed in a rectangular codex book form. Only the spiral scroll can symbolize the unending flow of divine light. Both form and medium are integral parts of the message. Changing the medium can change the message radically; a Torah scroll written on Japanese rice paper is bizarre and one written on pigskin would be the ultimate anti-Semitic statement.

The biblical object of supreme worth is the Ten Utterances engraved on stone tablets with divine laser light. What did Moses do with this "Made by God" object? He smashed it to bits. Did God reprimand or punish him for what would appear to be an immensely sacrilegious act? No. God tells Moses that he should carve out a second set of stone tablets and "write on the tablets the words that were on the first tablets *which you broke*" (*Deuteronomy 10: 2*).

In the Talmud, the rabbis ask why did God have to remind Moses that he had broken the "Made by God" tablets. He knew that he had smashed them. The words "which you broke" seem superfluous. One of the rabbis, Reish Lakish, plays with the Hebrew word "which" – *asher* – that can also be read as *ashrei*, "praiseworthy."[20] God praises Moses for the wisdom he had in breaking what was divinely engraved in stone to prevent our idolizing a frozen formulation. The Torah must be written by a human hand. We must create an intimate bond with it, rather than holding it in distant awe. I heard it told that the Hebrew letters from the smashed tablets spread out over the desert as an invitation for each generation to collect them and rewrite the Torah for itself in the spirit of the times. The divine original is replaced by human copies and copies of copies over generations to our day.

Postmodern theory often cites Walter Benjamin's seminal essay, "The Work of Art in the Age of Mechanical Reproduction."[21] He argues that reproduction emancipates a work of art from the ritual of honoring an original. Since one can make any number of prints from a photographic negative, to ask for the "authentic" print makes no sense. When the criterion of authenticity ceases to be applicable to artistic production, the total function of art is changed. It creates a political situation in which all members of the community can gain access rather than only the privileged and powerful few. The postdigital age extends Benjamin's observations on relationships between original and copy further, into situations in which there is no original at all, only invisible, immaterial bits and bytes posted on the Internet and materialized by anyone anywhere in the world.

Although the "Made by God" original needs to be smashed to bits to prevent it from being kept at a distance and idolized, only high touch Torah scrolls flawlessly copied by hand retain ritual power. Printed copies, widely disseminated throughout the community, are valued for Torah study, but not for public reading in synagogues. The essence of biblical study is to smash the text to bits and put it back together in multiple alternative ways. This postmodern-like deconstruction leads to reconstruction with its fresh insights and new perspectives gleaned from the Torah, a process called *hiddushei Torah*.

Indexic art: documentation

The third class of representational art is indexic. If we return to our example of a man walking on the beach, we see that a painting that looks like a man walking on the beach is iconic art, and that words painted out on a canvas are symbolic art. The actual footprints in the sand indicating that a man had walked on the beach can be perceived as indexic art. Indexic art represents occurrences by presenting direct physical evidence that they occurred. The word "index" is used as in its original derivation from Latin *indicare*, meaning to indicate, to point out as an index finger does. Indexic art documents events in real space and in cyberspace.

Unlike painting that obscures the brushstrokes that indicate the process of painting, Van Gogh's vigorous brushstrokes are indexical documentation of applying impasto paint with a paintbrush. He made his brushstrokes, evidence of process, an integral part of his artistic statement. Although indexical signs are felt strongly in Van Gogh's paintings, he continued to maintain iconicity in them. The full abandonment of the icon in painting and its replacement with pure index occurred most powerfully in action painting. A Jackson Pollock painting is indexic art that displays symptoms of the artist's having dripped paint, as well as a documentary map and after-the-act choreographic score of the movement of his body over a canvas floor. There is a direct physical connection between the artist dripping paint and the dripped paint on the canvas. Indexic art represents by correspondence, directly connecting what was to what is.

Photographs, at first impression, would seem to be the epitome of iconic art, the zenith of iconicity, since they represent the most accurate visual likeness of an object or event. On closer scrutiny, however, it becomes clear that the very high iconicity results from the photographic image being produced by point-to-point correspondence between light rays coming from what is being represented and a chemically or electronically sensitized plane. From this point of view, photographs are indexic art forms, documentary records produced by direct physical connection between what was and what is. Cinema and video are indexic forms that add motion and sound to photography. Indexic pictures that render the invisible visible play a vital role in contemporary science. The work of many scientists involves reading symptoms of natural occurrences from X-rays, MRIs, electrocardiograms, spectrograms, scintigrams, seismograms, voiceprints, and numerous other technologically generated indexic pictures.

Conceptual art, earthworks, site-specific installations, and performance art rely on photography, video, and film for documentation. Often these indexic records of time-based and temporary artworks are presented in galleries and museums and are viewed as the art itself. Art critic Carter Radcliff, in *Out of the Box: The Reinvention of Art*, traced the deliberate instability of these works as a reaction to the stability of the minimalist object.

> Performance pieces vanished as they were seen. Lines drawn on the desert floor disappeared more slowly, yet they too made the argument that works of art need not be permanent. It's enough for them to survive in their documentation, and sometimes a piece did not come fully into being until a record of it went on view [...] Documentation is primary: the work itself, not a reference to work elsewhere.[22]

Patricia Norvell recorded the thoughts, ideas, and feelings of the artists who were redefining art in 1969 by interviewing them. In response to the renewal of interest in conceptual art, the transcriptions of Norvell's interviews were published in 2001 in *Recording Conceptual Art*. One of her standard questions was, "How does documentation function in your work?" She found that the redefinition of art centered mainly on the role of the object in art. "For many artists the object as art is obsolete. In their work, either the object is eliminated entirely or, if it is employed, its formal elements are subordinate to such concerns as material quality, natural phenomena, natural forces, location, process or system. Where the art object is eliminated, some documentation of the art idea is usually substituted. Thus what is presented to the viewer may be photographs, written documentation and descriptions, or spoken information."[23]

A high tech example of indexic art is one of the *Four Wings of America* pieces I created in celebration of Miami's centennial in 1996. An electronic documentation was generated of the multiple branching pathways of a cyberangel traveling via the Internet between the four corners of America: Miami, San Diego, Seattle and Portland (Maine). With a scanner,

I dematerialized the Rembrandt drawing of the same angel that I had formerly sent around the globe by satellite. This time, however, the Internet transported the cyberangel, restoring the sense of geographical space lost when surfing the Web.

Did you notice how a web page that you are receiving does not appear all at once on your monitor? It comes up on your screen in parts until the whole finally comes together. The full cyberangel image does not fly out through the Web at once. The Web server sending the digitized image to the requesting browser breaks the image up into data packets. Each packet is assigned an ID number and routed by routers from one corner to the next through the available telecommunications pathways. Hence, the single image is deconstructed and routed through cyberspace between Miami and San Diego along multiple pathways. When the data packets reach San Diego, they are reassembled in the correct sequence based on the ID numbers that were assigned in Miami. The transmission control protocol (TCP) ensures that all the packets get to the requesting computer with no pieces missing as the whole cyberangel is rematerialized.

One angel packet can fly from Miami to New Orleans to Houston to Albuquerque to Phoenix to San Diego, while another angel packet flies from Miami to Atlanta to Nashville to St. Louis to Tulsa to Denver to Las Vegas to San Diego. Visualize the documentation of hundreds of routing paths plotted between the four corners on a map of the USA. The erratic pathways drawn from Miami to San Diego, from San Diego to Seattle, from Seattle to Portland, and from Portland back to Miami appear to possess electric energy. The indexic record of the cyberangel flight around the American perimeter looks like flashes of lightning. Cyberangel flight is the appropriate metaphor since the Hebrew word for corners, *kanfot*, as used in the biblical phrase "the four corners of the land" literally means "wings," like wings of birds and angels. Moreover, digital data packets in cyberspace provide a good analogy for angels themselves. In Jewish tradition, angels are discrete data packets in the immaterial Worlds of Creation and Formation realized in the everyday World of Action. An angel in the World of Creation is a one-of-a-kind cognitive data packet of a specific thought, word, idea or concept. An angel in the World of Formation is an affective data packet of a particular feeling or emotion, a specific inclination or impulse toward love, fear, pity, and so on. No two angels are alike.

The role of the angel is implicit in its Hebrew name, which means "messenger." Like data packets transporting information through cyberspace, the task of angels is to maintain two-way communications between our World of Action and Worlds of Formation and Creation. They are emissaries bringing divine plenty down into our material world and raising up certain consequences of human actions.[24]

In *High Tech High Touch: Technology and our Accelerated Search for Meaning*,[25] John Naisbitt proposes that each new technology elicits a compensatory human response, the more high tech, the more high touch. Whenever a new technology is introduced, there is a counterbalancing high touch human response or the high tech innovation is rejected. As more courses in computer graphics were offered in art schools and universities, there was a parallel demand for more courses in ceramics. Beyond the dual directions of high tech and high touch, high touch is beginning to inform high tech consciously integrating digital technologies into our lives. In keeping with Naisbett's formula of high tech/high touch complimentarity, like husband and wife, my geographical Webart is counterbalanced by geological clay art.

The artist Miriam Benjamin found inspiration from studying nature's documentation of the interaction of geological and environmental forces in the Negev desert mountains of Israel described by vivid biblical imagery: "The earth beholds and trembles. The mountains melt like wax." "The sea beheld and fled; the Jordan turns backwards; the mountains skipped like rams, the hills like lambs" (*Psalms 97, 114*). During the seven years she lived in the desert mountains, she developed an aesthetic vocabulary of earth forces. She wrote in her Pratt Institute MFA thesis:

> The desert is where shapes created by powerful geological forces from beneath and by erosion of wind and water from above are most apparent. The desert landscape is raw earth. It is not covered by grass, by trees, by snow, or by buildings and pavement [...] After flash floods in the desert, I watched the wet earth dry out and crack into beautiful patterns like the skin of a giraffe. On the hillsides, erosion wrinkles the earth like the skin of an aged elephant. Sometimes the fast-moving water leaves patterns like the feathery frost on a winter windowpane or like the venation patterns of a huge tropical leaf.[26]

She enlisted the aid of a geologist from the university in Beersheba to locate and dig clay from the mountains. The clay would become a documentary record of a dialogue between her hands and the malleable earth – the delicate, yet powerful, movements of her fingers beneath a slab of wet clay. Her pushing, pulling, lifting, tearing, pressing, scoring, folding and pinching causes the clay to crack, wrinkle, warp, rupture, slump, swell, shear, part, gnarl and burst. Sometimes, she smoothes areas into flowing, quiet plains.

The resulting *Clayscapes*, like Pollock's paintings, are indexic, documentations of creative interactions between an artist and the material world. Benjamin's work, however, reaches the deepest level of high touch. She feels the energies of earth in her bare hands, touching, holding, caressing flexible slabs of clay, listening to the clay tell her fingers what it can become. Paulus Berensohn, author of *Finding One's Way With Clay*, wrote: "The artist who most fills me with hope is Miriam Benjamin. She goes back to the mineral world in a way I celebrate. She is about healing, about going back to the desert, about listening."[27]

Identic art: being

Categories of representational art signify what *was* by illustration, symbolization and documentation. Presentational art forms signify what is, what can be, and what is becoming. Identic art does not look like something else, nor does it symbolize or indicate something other than itself. It is form and color presented as form and color; it is a real thing presented as itself, it is a real time electronic transmission of an event, and it can be an everyday event that is presented as life being lived. Alan Kaprow suggests that all of these identic trends evolved from Cubist explorations.

> Mondrian saw in Cubism the precursor to a nonfigurative, transcendent formal language. This lofty sense of abstraction continued to resonate through Newman and Reinhardt and well into Minimalism. In contrast, Duchamp picked up from that same Cubism's collages and constructions the ironic possibility that the artist's selective appropriation of commonplace materials and mass-produced images might replace the artist's traditional skill and individual creativity.[28]

Duchamp's Readymades led to Assemblage, Events, Earthworks, and Kaprow's own work from Happenings to ordinary life performed as art/non-art, lifelike art rather than artlike art.

Kandinsky aimed for a pure identic art in which form and color has an abstract presence like music that does not represent anything other than its own sounds. He was critical of Cubism "in which natural form is often forcibly subjected to geometrical construction, a process which tends to hamper the abstract by the concrete and spoil the concrete by the abstract."[29] He saw the danger that if we "devote ourselves purely to combinations of pure color and abstract form, we shall produce works which are mere decoration, which are suited to neckties and carpets."[30] In *Point and Line to Plane*,[31] his sequel to *Concerning the Spiritual in Art*, he describes the spiritual values of forms and colors that invite us to grasp their inner resonance, thereby effecting corresponding vibrations of the spirit.

Perhaps Mark Rothko's paintings of hazy-edged patches of color weightlessly floating in the center of large, tall canvases are powerful exemplars of Kandinsky's view that pure color and form can evoke spiritual responses. The atmosphere adduced by Rothko's pictures strike a deep chord in many viewers, evoking emotions and associations that have often been described in mystical, spiritual, or religious terms. 'Rothko: Art as Religious Faith' is how Hilton Kramer headlined a review in the *New York Times* of the major Rothko exhibition held at the Guggenheim Museum in 1978. Art Digest's staff reviewer referred to the vaporous, mystical Mark Rothko.

In her book *Mark Rothko: Study in Abstraction*, Anna Chave quotes from Dore Ashton that most of Rothko's friends agree that his "preoccupation with spiritual values was almost traditional." As a young child in Russia, his mildly religious and intensely Zionist family had sent him to *ẖeder* where he was given a traditional Jewish education. Although he became less observant as an adult in America, his roots, no doubt, colored the spiritual dimensions of his art.[32]

We can trace the development of identic painting discussed in Kandinsky's 1911 book *Concerning the Spiritual in Art*, to Kasmir Malevich's 1918 non-objective painting *White on White*, which presents a white square hovering diagonally on a white ground of a slightly different hue. The Dutch *de Stijl* movement further purified this identic direction in the 1920s.

Its leading proponent, Piet Mondrian, limited his artistic vocabulary to black and white and the primary pigment colors – red, blue and yellow – painted on a vertical and horizontal grid that seemed to open out into the space surrounding them. In America of the 1960s, Ad Reinhardt created a series of paintings in which black squares of slightly different hues gradually reached the limit of human ability to discern the difference between them through visual perception. Josef Albers, Kandinsky's younger colleague at the Bauhaus, painted his *Homage to the Square* series and published his *Interaction of Color* while heading the Yale Art School. Albers set solid colored areas in relationships that challenge our perception of color and space.

The extension of this identic formalism and minimalism from painting to sculpture can be seen in the work of Donald Judd. He ordered boxes from a sheet-metal shop and had them delivered to an art gallery. These commercially built sheet metal boxes were exhibited. Judd set them equidistant from one another to evoke awareness of boxes. He had no intention of representing anything, only presenting the boxes as themselves.[33]

Agnes Denes created a public art project, *Wheatfield, Battery Park City – A Confrontation*, which honors the tenacity of life against the encroaching city. Barbara Matilsky describes this monumental work of environmental art in *Fragile Ecologies: Contemporary Artists' Interpretations and Solutions*:

> On landfill near the World Trade Center in New York City, Denes cleared debris and garbage from an unused 4-acre parcel, brought in 225 truckloads of earth, and planted 2 acres of wheat in 1 inch of topsoil. With two assistants and some volunteers, the artist created an irrigation system and maintained the field for four months. During the summer, gleaming green stalks of wheat, which eventually turned glowing amber, were seen swaying against a fortress of skyscrapers.[34]

The cover of *Fragile Ecologies*, featuring a 1982 photograph of Denes' artwork looking up at the twin towers through "amber waves of grain," evoked memories of singing "America, America" with my fourth-grade class in New York. This tranquil life-affirming scene of America's agricultural and financial successes was brutally shattered by the horrific TV images of the twin towers falling in a fiery inferno. Life-hating militant Muslims had declared holy war against America's cherished values. Juxtaposed against constant replays of the hijacked plane crashing through the second tower, TV news showed Hollywood scenes of the apocalyptic destruction of the Empire State Building and a computer game won when a virtual World Trade Center and a virtual Pentagon are blown up. Life imitating art can be hideous.

Allan Kaprow contrasts artlike art with a lifelike art that is purely identic, identical with life's daily routines. Lifelike artists can be found playing at living life with awareness.

> They find life's meaning in picking a stray thread from someone's collar. And if that isn't it, they find in just making sure the dishes are washed, counting the knives, the forks, the

> cups and saucers as they pass from the left hand to the right. How different this is from "artlike artists," whose art resembles other art more than anything else. Artlike artists don't look for the meaning of life; they look for the meaning of art.[35]

Kaprow decided to pay attention to brushing his teeth alone in his bathroom, without art spectators. He brushed his teeth attentively for two weeks becoming aware of the tension in his elbow and fingers, the pressure on his gums, and the slight bleeding that made him think he should see a dentist. Kaprow describes the awakening of his awareness:

> I began to pay attention to how much this act of brushing my teeth had become routinized, nonconscious behavior, compared with my first efforts to do it when I was a child. I began to suspect that 99 percent of my daily life was just as routinized and unnoticed; that my mind was always somewhere else; and that the thousand signals my body was sending me each minute were ignored.[36]

In response to his question of the relevance of his brushing to art, he traces art's shift away from objects in a gallery to the real body and mind, communications technology, the real urban environment and natural regions of the ocean, sky and desert.

> Thus the relationship of the act of toothbrushing to recent art is clear and cannot be bypassed. This is where the paradox lies; an artist concerned with lifelike art is an artist who does and does not make art […] But ordinary life performed as art/not art can charge the everyday with metaphoric power.[37]

In *The Origin of Consciousness in the Breakdown of the Bicameral Mind*,[38] Princeton University psychologist Julian Jaynes explores the Hebrew Bible as a document tracing the development of human consciousness. His radical theory of the bicameral mind holds that people at the time of the patriarch Abraham could not think as we do today. They lived all their lives in an unconscious state. We know that state from routinely brushing our teeth or driving a car while talking to our passenger and listening to the radio. We do not pay full attention to the brushing or driving. But these ancient peoples had no way of reaching awareness by choice as we have today. They were not conscious of whether they were conscious or not. From the visual cues issued by the outsized eyes of their idols, they heard orders from gods, which were really one hemisphere of their brains speaking to the other one. Abraham abandoned the idolatry of his birthplace and his father's house and began the journey toward consciousness that was realized in the higher level of human awareness of the prophets. The Hebrew Bible is a history of the evolution of psychological awareness, freedom of thought, and moral consciousness.

Until this day, the Hebraic tradition celebrates awareness of our actions in our everyday life. We are enjoined not to pass through life without consciously knowing what we are doing. "Knowing" in Hebrew, *yada*, combines the words for hand and eye. Our eyes see what our hand is doing. We need to be aware of our actions and be morally responsible for them. Upon arising each morning, before getting out of bed, we declare: "I gratefully thank you, living and eternal king, for having returned my conscious soul to me with loving kindness." We then get up and do our first act of the day – washing our hands. Since hand washing, like tooth brushing, is usually an unconscious routine, a ritual was devised 2,000 years ago to make certain that the first act of the day is performed with full awareness. After our declaration of thanksgiving for reviving our awareness from the night's slumber, we take hold of a two-handled vessel filled with water and begin the hand-washing process. The set of instructions for this process found in the most commonly used Hebrew prayer book in America reads like a proposal for a conceptual art piece:

> Wash the hands according to the ritual procedure: pick up the vessel of water with the right hand, pass it to the left, and pour water over the right. Then with the right hand pour water over the left. Follow this procedure until water has been poured over each hand three times.[39]

Real time electronic transmission of events provides the greatest potential for creating identic postdigital art. Artists have the potential of employing such technologies as live television broadcasts, teleconferencing, and webcam image distribution. As part of my *Four Wings of America* collaboration with artist Miriam Benjamin, we created the framework for dancers from the four corners of America to electronically dance with each other in real time on a single screen although their real bodies were simultaneously active in Miami, San Diego, Seattle and Portland (Maine). With the cooperation of the ABC-TV network and Daniel Lewis, Dean of Dance at Miami's New World School of the Arts, and a National Endowment for the Arts grant, Miriam and I invited the heads of dance companies from the four corners cities to come to Miami to jointly choreograph a work celebrating Miami's centennial. Using up-link satellite transmission, dancers thousands of miles apart in different time zones could interact with each other in real time. A viewer could watch four groups of dancers move with and through each other on the same TV screen against a background of the interplay of Atlantic and Pacific Ocean waves.

The webcam, however, provides greater potential for artists than expensive TV broadcast technologies. In a *New York Times* article, "Seen My Sock Drawer Lately? Check Out My Web Site," Tom Zeller reports of the Ebay Internet auction of a non-working, waxy-white Krups ProAroma coffee machine. It had been leaking so the sellers cut off the electric cord. Although its original price was about 50 bucks, the broken coffee pot sold for just under $5,000. It seems the Trojan Room Coffee Pot, as the appliance is known, had been under the steady gaze of a video camera in the computer lab at Cambridge University in England. In 1991, enterprising workers set up the system and distributed three images per minute over the local network, to spare themselves the long hike to what might be an empty pot. Two years later, a new camera was mounted and the images were linked to a nascent, global network called the Internet, the world's first "Web cam" – an entire sub-genre of Internet culture – was born. People around the world could then check if there was coffee in the pot. The Ebay auction purchaser referred to the pot as a bit of a "digital joke," but pointed out that it is "one of the few true pieces of Internet history."[40]

Although the Coffee Cam has been switched off because the Cambridge lab is moving, we can monitor the growth of corn in an Iowa field in real time online at www.corncam.com; watch sharks, sea turtles and barracudas swimming in the four-story ocean tank at the New England Aquarium in Boston, or see crowds gathering for the hourly mechanical parade of saints in the clock tower in Prague's historic city center, or hundreds of other real time events worldwide at www.earthcam.com.

28,809,416 people have experienced the Western Wall (*kotel*) in Jerusalem via a webcam at www.aish.com/w/ from when it was activated until the Purim holiday during the year 5770 of the Hebrew calendar corresponding to the first of March 2010. True to Jewish form, this passive observation of the Wall can be transformed into a participatory experience with the click of a mouse. Viewers are invited to click on four options: placing a note in the wall, dedicating the webcam image to the memory of a loved one, listening to spiritual music, or emailing the webcam page to a friend. Viewers are invited to use the online form provided. "It is a centuries-old tradition to place a note with a prayer or request in the Western Wall. Type in your prayer. It will be printed out in the Old City of Jerusalem where it will be placed in the Wall by a student of Aish HaTorah" From an alternative vantage point we can see live video feed from the Western Wall at the www.virtualjerusalem.com/kotelcam.php. At this website, viewers can participate in using control technology to move the webcams. Relatives and friends worldwide can actively participate in a real time bar mitzvah celebration at the Wall by zooming in and out and panning to the left and right. This site also offers to have its staffers bring emailed prayers to the *kotel* if they are fewer than 200 characters.

At both *kotel* websites, we can watch people writing their prayers, hopes, and aspirations on little slips of paper, which they push into the tight spaces between the massive stones of the remaining wall of biblical Temple times. It is instructive to note that Jews interact, not with the mighty, impressive, rectangular stones, but with the open spaces between them.

##Prioric art: proposing

Iconic art, symbolic art, and indexic art are past oriented; identitic art exists in the present; prioric art is future directed. Prioric art is the presentation of a proposal or plan for a potential event, an *a priori* statement of what can be. It often employs iconic and symbolic modes of signification for presenting itself. The prioric form is more common in art forms other than the visual arts.

It can take the form of scores in music and dance, scripts in theater and film, or architectural plans. Like these forms, the visual artists can propose artworks that they do not make themselves. Musicians perform music created by composers, dancers move to choreographers' notations, actors enact a script written by playwrights, and building contractors convert architectural drawings into buildings.

Visual artists act more like composers, choreographers, playwrights, and architects in creating prioric art. The proposal or plan can be presented as the work of art itself. John Cage has exhibited his musical scores as works of visual art. One of his scores is presented as a standard-looking title page of sheet music from without, but within it is blank with the exception of a few words printed on it: "A work of piano consisting of total silence for four minutes and thirty-three seconds. The work, however, may be performed by any instrumentalists or combination of instrumentalists." Architectural drawings and models, although plans for building a building, are often sold in art galleries as works of art.

Contemporary visual artists create prioric art as proposals for art world artworks – plans for public art installations, scores for performance art, and proposals for art exhibitions. Visual artists can also create prioric art as proposals, scores, and plans for political, social, or ecological action in the arena of real life. In all these cases, the prioric art can be presented as an artwork in itself even if it is never realized.

Prioric art as a complete artwork in itself is most highly developed in the work of the husband-wife team, Christo and Jeanne Claude. All of their major environmental art installations have been financed by prioric art created by Christo and marketed by Jeanne Claude in the years before the realization of their projects. They present plans for their projects as preparatory drawings, collages, and scale models, which are sold directly to museums and collectors. They never accept sponsorship of any kind. Their prioric artworks are both proposals and the source for funding installations that can cost a million dollars.

Christo began making drawings and collages in 1979 proposing the installation of 7,532 gates over the 23 miles of pedestrian walkways in New York City's Central Park. His prioric works show free hanging saffron colored fabric panels suspended from the horizontal top part of the gates. The gates are spaced at intervals that allow the synthetic woven panels to wave horizontally towards the next gate and be seen from far away through the leafless branches of the trees. In 2003, they received permission to realize their temporary work of art, *The Gates*. His proposals for the project – 45 drawings and collages, 40 photographs, and 10 maps and technical diagrams were exhibited at the Metropolitan Museum of Art prior to the project's 16-day realization in the adjacent Central Park in 2005.

Christo considers *Surrounded Islands* their most beautiful artwork looking like giant pink water lilies floating on the tropical blue-green waters of Biscayne Bay. A preview of its beauty was seen in his preparatory pastel and charcoal drawings, collages and lithographs, the prioric art that they sold to pay for 6.5 million square feet of pink woven fabric surrounding eleven islands. They needed a work force of 430 people to unfurl the pink fabric and 120 monitors in inflatable boats to tend the giant blossoms by day and night.

> For two weeks *Surrounded Islands* spreading over 7 miles was seen, approached and enjoyed by the public, from the causeways, the land, the water and the air. The luminous pink color of the shiny fabric was in harmony with the tropical vegetation of the

> uninhabited verdant islands, the light of the Miami sky and the colors of the shallow waters of Biscayne Bay.[41]

It took many years from the creation of the prioric artwork to the realization of the project. Numerous bureaucratic hurdles had to be jumped before permissions from the different cities of Miami, North Miami, Miami Shores, and Miami Beach could be obtained.

Christo made his preparatory drawings and collages for *Wrapped Reichstag* 24 years before he actually wrapped the entire building. He started to seek permissions from the occupying American, British, and Soviet authorities as well as the local municipal bureaucrats. In the end, Berlin was united, and he only had to convince a single governing body. Sometimes his proposals never move beyond being prioric art. Christo's lithograph showing the Whitney Museum of American Art covered and tied with cloth and rope is a proposal for wrapping the museum that remains unrealized.

Lawrence Halprin created what he calls "scores." A score is a prioric art form that facilitates the interaction of individuals in a group with each other and their environment. Halprin prepared one of his scores by superimposing a calendar grid for the month of September on a map of San Francisco. He proposes that the performers of his score create events in the geographical area corresponding to a particular day in the calendar's grid. For example, if the calendar box marked with the number "29" falls over a portion of the map marked "Ocean View Park," then the performers enact their September 29 events in Ocean View Park. Each of the other day's events is planned to take place in another part of the city determined by the calendar/map matrix.

> What differentiates scoring from other arts ultimately is that scoring is a means to make other people creative. The measure of the success of a score is how much it achieves for others – not how creative or beautiful or amusing the score is in itself. Scoring, therefore, is an art form devoted to sharing and participation, and the scoring artist is an energizer and catalyst for collective creativity.[42]

Mierle Laderman Ukeles' "Maintenance Art Manifesto, Proposal for an Exhibition, 'CARE'" is published in full in Albero and Stimson's critical anthology *Conceptual Art*. Ukeles proposes bringing her everyday life into the art-world milieu as art. She quotes the Balinese saying, "We have no Art, we try to do everything well." As an artist exploring the revolutionary implications of art being redefined, she inquires, "The sourball of every revolution: after the revolution, who's going to pick up the garbage on Monday morning?" As the daughter and sister of Orthodox rabbis, she is talking *takhlis*, a Yiddish word that means, "Don't talk theory – what are you going to do?" Just as the Jew locates holiness in creatively encountering everyday life, Ukeles transforms her life as a housewife into art. She presents her proposal:

> I am an artist. I am a woman. I am a wife. I am a mother. (Random order.) I do a hell of a lot of washing, cleaning, cooking, renewing, supporting, preserving, etc. Also (up to now

> separately) I "do" Art. Now I will simply do these maintenance everyday things, and flush them up to consciousness, exhibit them, as Art. I will live in the museum as I customarily do at home with my husband and my baby, for the duration of the exhibition [...] and do all these things as public Art activities: I will sweep and wax the floors, dust everything, wash the walls (i.e. "floor paintings, dust works, soapsculpture, wall-paintings"), cook, invite people to eat, make agglomerations and dispositions of all functional refuse. The exhibition area might look "empty" of art, but it will be maintained in full public view. MY WORKING WILL BE THE WORK.[43]

Allan Kaprow pioneered in creating "Happenings". He was the first to use the word "Happenings" to describe performance art that invited scored participation of the audience in interaction with surroundings created by the artist. In the early 1950s, he was exhibiting his paintings and teaching art history at Rutgers University. Following his interest in Jackson Pollock's action painting, Kaprow's paintings became "action-collages" in which great hunks of varied matter protruded from his bigger and bigger paintings until the entire gallery space was filled up. "The materials were varied: sheets of plastic, crumpled-up cellophane, tangles of Scotch tape, sections of daubed enamel and pieces of colored cloth hung in bands that looked like Jewish prayer shawls or other ceremonial adornments."[44] As he watched visitors moving through the environments he had created, he began presenting them with scripts he wrote as proposals for their actively encountering his constructions. He wrote in his artist's statement in Michael Kirby's book, *Happenings*:

> I immediately saw that every visitor to the Environment was part of it. I had not really thought of it before. And so I gave occupations like moving something, turning switches on – just a few things. Increasingly during 1957 and 1958, this suggested a more "scored" responsibility for the visitor. I offered him more and more to do, until there developed the Happening.[45]

Since he wanted people involved rather than as spectators, he wrote down on a sheet of paper actions that could be learned by anyone. These scripts were given to those people who wished to participate in advance of the Happening. "Thus, when I arrived shortly before the scheduled event, I already had a committed group, and I could then discuss the deeper implications of the Happening with them as well as the details of performance."[46]

The precedent for the Happening was established in Jewish tradition millennia ago. It is the Passover *Seder*. Unlike Greek theater that developed into proscenium theater with actors playing to a passive audience, the Happening and the *Seder* draw their actors from the audience. A closed script guides the actor who plays Hamlet. He follows Shakespeare's exact words in playing a role before spectators. Open-ended scores that invite creative dialogue with others and with a prepared environment guide participants in a Happenings or a *Seder*. When I discussed the influence of the *Seder* on the creation of the Happening with Kaprow, he explained how his work evolved within an art framework from painting, to action-collage, to environment, to Happening. However, he did acknowledge his participation in his family *Seder* on Passover that made him subconsciously aware of the structure of this traditional Jewish form of participatory performance art.

The aim of the *Seder* is for all participants to feel as if they themselves had been freed from bondage in Egypt. An environment for multi-sense involvement is set to evoke questioning. Participants read a *Haggadah*, an open-ended score to guide them in their interaction with a specially set table that evokes the telling the story of the exodus. It is a multifaceted score that begins with questions asked by the youngest child followed by sources that encourage animated discussion and group singing by all, from the youngest child to the aged sage.

Eating and drinking are designed into the score to be an integral part of the story-telling experience. Bitter herbs are eaten until tears well up in our eyes to feel the bitterness of slavery. A brown mush looking like the mortar that the slaves were forced to form into bricks is eaten. It is made from ground apples and nuts mixed with sweet red wine and honey. Its sweet taste tells us how we can be fooled into becoming slaves to empty values. This mushy *haroset* is eaten with crunchy *matzah*, the bread of affliction and poverty that had no time to rise in our haste to leave Egypt. Both flat, round hand-made *matzah* and rectangular machine-made *matzah* are eaten. The circle symbolizes idolatry and the rectangle slavery in contrast to the spiral *h̲alah* eaten after Passover every Sabbath to symbolize freedom and organic growth. Children scramble in search of a hidden piece of *matzah* required as the desert after a full meal eaten in the middle of the *Seder*. Drinking four cups of wine during the duration of the *Seder* symbolizes stages toward redemption by relaxing the participants into feeling their freedom while reclining on puffy pillows.

The prioric artwork of Christo and Jeanne Claude are drawings, collages, and scale models as plans for environmental public art installations. Halprin's scores are plans for performance art pieces that relate time and space in new ways. Ukeles' proposal brings the everyday world into the art world through the artist's performance in a museum setting. Kaprow's scores direct participants who are not artists to interact with an art environment that he created. The structure of Kaprow's Happening resembles the Passover *Seder*, an ancient ritual event that anticipates modern performance art. The *Haggadah* as an open-ended score for intergenerational, multi-sense involvement is a model for creating postmodern prioric art. Contemporary artists are also creating artworks that are proposals and plans that act as catalysts for effecting real change in the real world. Exemplary prioric artworks acting beyond the limitations of the art world are Helen Mayer Harrison and Newton Harrison's ecological artwork *Breathing Space for the Sava River* and my political piece *Aesthetic Peace Plan for the Middle East.*

When the husband-wife team of ecological artists Helen Mayer Harrision and Newton Harrison were in Berlin, they learned about the Sava River ecosystem in Yugoslavia, one of Europe's last great flood plains. Although the river is rich with wildlife and many farmers are engaged in traditional sustainable agriculture, industrial wastes pollute the river confined in canals and dams built for irrigation and flood control. The Harrisons photographed the course of the Sava River from its twin sources in mountain and swamp before it empties into the Danube River. The river is pristine at its beginning where rare black storks, spoonbills, and sea eagles can be seen. Further downstream, it becomes poisoned by the outflow of waste products from a nuclear power plant and factories. The artists documented the toxins excreted into the river by a paper mill and fertilizer plant.

In an effort to reverse this ecological disaster, the Harrisons created an artwork as a plan for restoring the Sava River. It included photographs and poetic texts to illustrate the collision between man and nature along the river. They proposed planting carefully selected plants that could form a natural root-zone purification system to eliminate many pollutants. Organic farming along the edges of the preserve was recommended as a method of reducing fertilizer runoff and stem destructive algae bloom. The water used to cool the nuclear power plant could be recycled into holding ponds for raising warm-water fish.

After exhibitions of their prioric artwork, *Breathing Space for the Sava River*, in Germany and Yugoslavia, the Croatian Department of the Environment approved the plans and the World Bank expressed interest in funding the river cleanup. Wherever it has been exhibited, the Harrisons' ecological artwork has simulated debate that has lead to positive actions. In the catalog for the exhibition *Fragile Ecologies: Contemporary Artists' Interpretations and Solutions*, Barbara Matilsky writes:

> Increasingly, communities and governments are responding by adapting the Harrisons' ideas into their own long-range plans for addressing the quality of the environment. The Harrisons are also defining the future direction of ecological art by expanding the role of artists who, it is hoped, will become members of teams that implement projects relating to large ecosystems.[47]

My prioric artwork *Cyberangels: Aesthetic Peace Plan for the Middle East* proposes a solution to a problem in international relations that has been festering for over a century. It proposes that the Arab conflict with Israel is an aesthetic problem that calls for an artistic solution. It calls for a shift in perception derived from Islamic art and thought. In the catalog for the presentation of this artwork in a 2004 exhibition at the Robert Guttmann Gallery of the Jewish Museum in Prague, the museum's curator of contemporary art, Michaela Hajkova, explains:

> Alexenberg has given Rembrandt angels, symbols of European culture, the role of peacemakers and mediators, disseminating across the Internet a message that calls for a paradigm shift in the way the Middle East conflict is perceived. Alexenberg has focused on a purely visual concept – the relationship between pattern and counter-pattern, as demonstrated by the example of Islamic *kilims* – interpreted as a viable model for promoting mutual respect and recognition of the distinctness of Arabs and Jews.[48]

In my aesthetic peace plan, human creativity at its best in both Islamic and European cultures encounter each other. Beautiful patterns of Islamic art meet Rembrandt's angels in conveying a message of peace. Indeed, the Hebrew term for angel of peace, *malakh shalom*, differs little from the Arabic term, *malak salam*. Digitized Rembrandt angels emerging from Islamic geometries are electronic age messengers drawing out the beauty of European and Islamic cultures rather than the ugly anti-Semitism that plagues them. My artwork invites a perceptual shift through which Muslims see the State of Israel as a blessing expressing Allah's will and Christians see it as the divine fulfillment of the biblical promise of the Land of Israel to the Jewish people.

To demonstrate perceptual shift, I showed my serigraph of a highly digitized detail from the Rembrandt etching showing an angel stopping Abraham's murderous act. The pixels are so blatantly large that one sees an abstract picture of the geometrical organization of squares. Only when the viewer stands back far enough to blur the pixels does the image of the angel and Abraham, the father of both Isaac and Ishmael, clearly appear. The change from an abstract image to a figurative one only exists in the mind of the viewer.

Historian of Islamic art, Elisabeth Siddiqui, writes that art is the mirror of a culture and its worldview. She emphasizes that there is no case to which this statement more directly applies than to the art of the Islamic world. "Not only does its art reflect its cultural values, but even more importantly, the way in which its adherents, the Muslims, view the spiritual realm, the universe, life, and the relationships of the parts to the whole."[49]

Islamic art teaches Arabs to see their world as a continuous geometric pattern that extends across North Africa and the Middle East. Unfortunately, they see Israel as a blemish that disrupts the pattern. From this perspective, Israel is viewed as an alien presence that they have continually tried to eliminate through war, terrorism, and political action. Arab television calls Israel a "cancer in the body of the Arab nation." Iranian President Ahmedinejad announced to the world media that nuclear weapons that Iran is developing could "wipe Israel off the map" and in the words of his predecessor Rafsanjani one atomic bomb could remove the "extraneous matter" called Israel from the heart of the Islamic world. The major obstacle to peace is Islamic leaders' refusal to tolerate a tiny sovereign Jewish state in the midst of an immense Islamic world that runs from the Atlantic Ocean to the borders of India.

Fortunately, the perceptual shift needed to lead to a genuine peace can be found in Islamic art and thought. In Islamic art, a uniform geometric pattern is purposely disrupted by the introduction of a counter-pattern that demonstrates that human creation is less than perfect. Based upon the belief that only Allah creates perfection, rug weavers from Islamic lands intentionally weave a patch of a dissimilar pattern to break the symmetry of their rugs to demonstrate that they are not competing with Allah. Sheikh Abdul Hadi Palazzi, Imam of the Italian Muslim community who holds a Ph.D. in Islamic Sciences by decree of the Saudi Grand Mufti, writes, "The idea of underlying the Divine infinitude and the human fallacy by including some 'voluntary defects' in works of art is common in Islamic art, and extends to tapestry, painting, music, architecture, etc."[50]

In *Islamic Textile Art: Anomalies in Kilims*, Muhammad Thompson and Nasima Begum write that weavers of Moroccan kilim rugs, "devout Muslim women would not be so arrogant as to even attempt a 'perfect kilim,' since such perfection belonged only to Allah. Consequently, they would deliberately break the kilim's patterning as a mark of their humility."[51]

In the exhibition, I showed actual kilim rugs that a Prague rug merchant loaned to the museum. I had photographs of the counter-patterns in these rugs enlarged digitally and printed out on canvas. These enlarged images of the anomalies were displayed hanging next to the actual rugs.

This aesthetic concept parallels deviations from symmetrical patterns that characterize life itself. All living organisms exhibit the principle expressed by the renowned biologist Paul Weiss as "order in the gross with freedom of excursion in details."[52] To demonstrate this principle, I exhibited a cross-section of a pine leaf that I enlarged 600 times by photographing it through a microscope. I mounted the photograph on a shaped panel and painted with encaustic paints. On first glance, the cellular organization in the cross-section gives the appearance of bilateral symmetry. On closer examination of the details, however, a quite different pattern could be discerned in each half of the leaf cross-section.

Peace can be achieved when the Islamic world recognizes that they need Israel to realize their Islamic religious values. Israel provides the counter-pattern in the contiguous Islamic world that extends from Morocco to Pakistan. Just as a religious Muslim weaver introduces a counter-pattern in designing a rug as a mark of humility, so Muslim leaders can exhibit humility while honoring the diversity in all of God's creations by perceiving Israel as the necessary break in symmetry. The ingathering of the Jewish people into its historic homeland in the midst of the Islamic world is the fulfillment of Mohammed's prophecy in the Koran (*Sura 17:104*): "And we said to the Children of Israel, 'scatter and live all over the world [...] and when the end of the world is near we will gather you again into the Promised Land."

Sheikh Palazzi quotes from the Koran (*Sura 5:20–21*) to support the Arab world's need to switch their viewpoint to recognize the sovereign right of the Jews over the Land of Israel as the will of Allah:

> Remember when Moses said to his people: "O my people, call in remembrance the favor of God unto you, when he produced prophets among you, made you kings, and gave to you what He had not given to any other among the people. O my people, enter the Holy Land which God has assigned unto you, and then turn not back ignominiously, for then will ye be overthrown, to your own ruin."

I used Photoshop to illuminate this Koranic verse attesting to the Jewish ownership of the Land of Israel with a pattern from an Iranian mosque. I colored the pattern in shades of Islamic green and digitally printed it out on canvas.

According to the Imam, Islam's holiest book confirms what every Jew and Christian who honors the Bible knows: The Land of Israel was divinely deeded to the Children of Israel. Jews are the indigenous people of the Land of Israel who have continuously lived there for more than three millennia despite the conquests of numerous imperialist empires.

Jews are from Judea. Arabs are from Arabia. The Arabs are blessed with 22 other countries. Enlightened Muslim leaders understand this. Tashibiah Sayyed, editor-in-chief of *Muslim World Today* writes:

> I consider the creation of the Jewish State as a blessing for the Muslims. Israel has provided us an opportunity to show the world the Jewish state of mind in action, a mind that yearns to be free [...] The Jewish traditions and culture of pluralism, debate, acceptance of dissention and difference of opinion manifest themselves in the shape of the State of Israel to present the oppressed Muslim world with a paradigm to emulate.[53]

The central image in the Prague exhibition was a large digital print on canvas derived from my computer-generated acrylic painting showing a geometric pattern from a Damascus mosque superimposed on a map of North Africa and the Middle East. The Islamic pattern is painted in green, red, and metallic gold. Israel is painted as a tiny sliver with an alternative pattern – blue stripes on a white background like the Jewish prayer shawl and the flag of Israel. An "angel of peace" derived from a digitized version of a Rembrandt drawing is shown emerging from the blue stripes carrying the message of my aesthetic peace plan for all to see.

I e-mailed this image with a descriptive verbal message in Arabic, Hebrew, and English to all the world leaders from the museum in Prague. Visitors to the museum could access responses using a computer keyboard linked to a large plasma screen that also showed my montages of digitized angels emerging from Prague subway posters. Since the image of biblical angels ascending on Jacob's ladder teaches that spiritual messages arise from below to bring peace, it seemed to me that they arise from the Prague underground.

This prioric artwork proposes that peace can come from a fresh metaphor in which the Arabs see Israel's existence as Allah's will. A shift in viewpoint where Israel is perceived as a blessing, as the necessary counter-pattern in the overall pattern of the Islamic world, will usher in an era of peace. This paradigm shift can transform the perception of Israel as a blemish to seeing it as a small golden seed from which a lush green Islamic tree has germinated, growing branches over North Africa and the Middle East.

The postdigital age of social networks has enabled the aesthetic peace plan to move far beyond the gallery walls eliciting participation of people worldwide in promoting peace in the Middle East. I created an Aesthetic Peace Facebook group with hundreds of members and an Aesthetic Peace blog (www.aestheticpeace.blogspot.com) which reaches out to thousands. Facebook description: "The Aesthetic Peace group aims through art and creativity to inspire new approaches and create new metaphors for peace between Israel and its neighbors." Blog description: "AESTHETIC PEACE. An aesthetic metaphor derived from Islamic art invites a perceptual shift through which Muslims see the Jewish State of Israel as a blessing expressing Allah's will rather than as an alien presence in the midst of the Islamic world. Perhaps thinking out of the box from a fresh aesthetic viewpoint can succeed in bringing peace where politics has failed."

Dialogic art: interacting

Dialogic art comes into being through dialogue. It exists as the interrelationship between people, and between people and their environments – natural, man-made and virtual. It can extend to inter-species dialogue. The difference between identic and dialogic forms of art can be described by Martin Buber's two primary words: I-It and I-Thou. I-It is the experience of something; it describes identic art. I-Thou, however, is not the experience *of* something, but rather an interrelationship that has its own existence. I-Thou comes into being through dialogue, the interactive shared sphere between people, a sphere of spiritual intensity.

> The participation of both partners is in principle indispensable to this sphere [...] The unfolding of this sphere Buber calls "the dialogical." The psychological, that which happens to the souls of each, is only the secret accompaniment to the dialogue. The meaning of this dialogue is found in neither one nor the other of the partners, nor in both taken together, but in their interchange.[54]

Dialogue between I and the human Thou extends to the partnership between I and the eternal Thou. The concept of the dialogic sphere between a unique human being and God in the human quest to complete God's creation, Buber regards as the central contribution of Judaism to the world's religions. The basic doctrine which fills the Hebrew Bible is that our life is a dialogue between the above and the below.

Rather than the Hellenistic concept of art as imitating nature by creating images and objects, the Hebraic concept of art is engaging in dialogue to create spiritual significance in our everyday lives. A powerful example of this dialogic art derived from the deep structure of Jewish consciousness is Mierle Laderman Ukeles' interaction with New York City sanitation workers. She spent eleven months shaking the hands of 8,500 men who collect garbage on the streets of New York. In Linda Montano's interview with her in *Performance Artists Talking in the Eighties*, Ukeles describes her dialogic artwork:

> It involved facing each one bodily and saying, "Thank you for keeping New York City alive." It was a ritual and a discipline for myself because I intended to mean exactly that as I faced and spoke to thousands and thousands of people. That was my own private goal: to watch myself very closely so that I wouldn't turn it into a mechanical thing. And I didn't. I proved to myself that I had ritual strength. It was not even hard because there was so much response from those people that I met, and that carried me. It was a very positive energy field that I was in.[55]

She discusses her learning to engage in ritual that continually renews itself from her father and brother, both Orthodox rabbis, and from her mother who set her free to engage in creative play.

Dialogic art sets *Vision in Motion*, to use the title of the book written by Kandinsky's Bauhaus colleague Laszlo Moholy-Nagy. It breaks down the duality between object and subject, between producer and consumer, between inside and outside, and between time and space. These dyads are grasped simultaneously. "Simultaneous grasp," Moholy-Nagy writes, "is creative performance – seeing, feeling and thinking in relationship and not in a series of isolated phenomena. It instantaneously integrates and transmutes single elements into a coherent whole."[56]

In his 1980 solo exhibition, *Beyond the Visible*, at the Guggenheim Museum in New York, Yaacov Agam included two types of art works to evoke dialogue: "polymorphic" and "transformable." In the polymorphic works, one interacts with the work by moving oneself; in the transformable works, one interacts with the works by moving parts of them. The polymorphic works are paintings on a series of vertical, attached triangular prisms. As the participant in the dialogue moves from the right to the left side of the prism, the painted forms on the right side disappear as the colors and shapes on the left side gradually become visible. In contrast with Cubism, where the artist experiences the many perspectives of reality and unifies them in a single image, Agam creates art works that invite each participant to unify the enormous number of perspectives that can only come into being through dialogic interaction. One of the two major polymorphic pieces in the Guggenheim show, *Panoramagam*, was attached to the interior wall spiraling 360 degrees around the museum atrium. The other is a 65-foot polymorphic tower, *Aenaitral Tower*, reaching from the floor of the atrium up to the ceiling. As we walk up the ramp, we continually see the changing facets of the vertical tower and spiraling polymorph from different viewpoints, with each view presenting a different image. His transformable, *Beating Beating Heart*, is a nine-element stainless steel floor sculpture of heart forms. When touched, each element moves independently. When all nine are set in motion simultaneously, a waving movement of heartbeat is established. Agam, son of a rabbi, states:

> A work of art which captures a specific moment and eternalizes it in a painting or sculpture is expressing a static view of existence. Authentic Jewish art must capture and communicate the very dynamism of life's flowing, changing quality [...] characterized by diversity, newness, aliveness, activity.[57]

Artist Jean Duppy collaborated with engineer Ralph Martel to create *Heart Beats Dust*, a biofeedback artwork shown at MoMA in New York. They constructed a glass-faced cube having a rubber-sheet floor covered with a fine powder that was strongly illuminated. Below the floor, an amplification system activated the powder in time with the rhythm of the participant's heartbeat sensed through a stethoscope. As she holds the stethoscope to her chest, she watches her own heartbeat visually displayed as an elegant dance of glowing powder leaping from the vibrating rubber sheet. She sees her inner pulse externalized in a rhythmic dialogue.

Please Superimpose, Please? is a video dialogue created by John Randolph Carter for an exhibition at the Whitney Museum of American Art. The artist invited two strangers who never saw each other in real space to engage each other in real time dialogue through electronic space. Carter placed a graceful young woman and a handsome young man on opposite sides of a wall. Although they could hear each other's voices, they could only see each other on the monitor on their side of the wall. Each one could see both himself or herself and the other person on a single screen. These strangers became acquainted by touching the air in such a way that their video-image hands touched each other. Their proxy touching grew more intimate leading to ghostly embraces and split-screen lovemaking. Virtual intimacy was not threatening. A new vocabulary evolved from their electronic interaction. She invited him, "Please superimpose your body on mine." The immediate feedback of video coupled with the diaphanous quality of two video bodies on a single monitor permitted their virtual bodies to merge in erotic dialogue that could only take place through an electronic medium as mediator. The couple flowed in and out of each other's translucent bodies moving simultaneously inside and outside of the TV box.

Agam and Duppy/Martel created dialogic art in which a person interacts with material and electronic environments. Carter created a dialogue between people with technology as mediator. Eduardo Kac collaborated with Ikuo Nakamura in creating *Essay Concerning Human Understanding*, an artwork of inter-species dialogue in which a canary dialogues over a regular phone line with a plant 600 miles away. At the Center for Contemporary Art, University of Kentucky, a canary lived in a cage on top of which circuit boards, a speaker, and a microphone wired to the telephone system were located. In New York, an electrode was placed on the leaf of a philodendron to sense its response to the singing of the bird. The voltage fluctuation of the plant was monitored through a Macintosh running Interactive Brain-Wave Analyzer (IBVA) software that played out sounds controlled by a MIDI sequencer. The order and duration of the sounds were determined in real time by the plant's response to the canary's songs. Interaction with humans standing near the bird and plant altered their behavior. The artists explain: "By enabling an isolated and caged animal to a have telematic conversation with a member of another species, this installation dramatized the role of telecommunications in our own longing for interaction, our desire to reach out and stay in touch. This interactive installation is ultimately about human isolation and loneliness, and about the very possibility of communication."[58] Kac told me that the spiritual dimension of his art is "searching for a balance between emotion and cognition."[59]

At the gravesite of the greatest kabbalist of the past millennium, Rabbi Issac Luria, better known as the Arizal, I copied the eulogy written by his student, Chaim Vital. Inscribed in Hebrew on a marble slab standing on a windswept hillside on the edge of the Galilee town of Safed, I read that Rabbi Luria was filled with Torah like a pomegranate and that he was expert in conversing with birds and angels.

At the *Software* exhibition at the Jewish Museum in New York, Nicholas Negroponte and the MIT Architecture Machine Group created an environment for dialogue between gerbils and a cybernetic system, which can give us clues about human dialogue with an increasingly computer-controlled world. Negroponte and his collaborators designed a habitat for a colony of gerbils to live among stacks of wooden blocks controlled by computerized pressure-sensing devices. Moving about in a 5 x 8-foot glass-enclosed, rectilinear environment, the gerbils moved about living their lives, bumping and pushing blocks here and there and sometimes knocking over an entire stack. A computer program, *Seek*, could arrange the blocks that were connected to a supporting carriage. As the blocks were moved around by gerbil activity, *Seek*, guided by its memory of the original configuration, continually attempted to re-establish the rectilinear pattern. Watching this ecosystem of life forces interacting with a technological environment makes us ask if we can extrapolate from this work of dialogic art, life in a city of the future. We can also ask: "Can new directions in media design be derived from our complex living patterns in a postmodern metropolis?"[60]

Artists who create interactive artworks using computer systems need to consider relationships between content and interface. Manovich proposes that old dichotomies *content-form* and *content-medium* have become *content-interface* in dialogic e-art. We are all familiar with mouse, keyboard, touch screen, or joystick as common interfaces. Through the interface we hold a dialogue with the content of the computer software. In fine art of the past, the forms and media selected by the artist are related to the content of the artwork. In new media art, the choice of interface, rather than form and medium, is motivated by the artwork's content. The interface becomes an integral part of the expression of content. Content and interface merge into a single entity and cannot be perceived as independent of each other.[61]

At the 2000 Knowledge Based Engineering Systems Conference, artists Laurent Mignonneau and Christa Sommerer presented a seminal paper, "Designing Interfaces for Interactive Artworks." They describe the invention of "natural interfaces" for their dialogic artworks. Natural interfaces can transport the content of life, variation and personality when the user interacts with a virtual space. Mignonneau and Sommerer's *Interactive Plant Growth* is an artwork in which the interface between the user and the computer is a living plant. Approaching or touching the plant wired to the computer triggers reactions in a virtual world of simulated plant growth. In their interactive computer installation *A-Volve*, visitors interact with artificial creatures that live, mate, and evolve in a water-filled glass pool. The creatures not only interact with each other but also react to the visitor's hands in the water. The artists installed a camera detection system that measured the visitor's hand positions and communicated the data to the artificial creatures. The creatures react to the visitor's gestures; they can stop moving when caught by the visitor's hand or act afraid when touched too often. In *Phototropy*, a light detection system measures the position and intensity of a spot of light from a flashlight on a large projection screen. As the user moves the light spot onto different parts of the screen, virtual insects appear and follow the light beam. The user can feed the insects with light or kill them by providing too much of it.[62]

Artists are continually inventing alternative interfaces that are conceptually and experientially linked to content in dialogic art systems. I developed the use of biofeedback interfaces in a dialogic artwork that I designed in collaboration with my son Ari at MIT's Center for Advanced Visual Studies. For the *LightsOROT* exhibition at Yeshiva University

Museum in New York, we used a plethysmograph interface, which monitors blood flow as an indication of changing mind/body states. Our bioimaging system integrates real time computer graphics with biofeedback.

The starting point for this dialogic artwork is the fact that the Hebrew words for face, *panim*, and inside, *p'nim*, are written with the same four letters. We explored creating a digital portrait in which outside flows from inside and inside flows from outside in a continuously flowing feedback loop. In our studio/laboratory at MIT we developed a system using biofeedback from brain waves sensed by electrodes connecting the participant's head to an electroencephalograph. For the museum, however, difficulty placing electrodes on people's heads required that we redesign the system. We built a console in which a participant seated in front of a monitor would place a finger in a plethysmograph, which measures internal body states by monitoring blood flow, while under the gaze of a video camera.

> A feedback loop is created in which changes in one's internal mind/body state changes a video image of one's external self. It is a video/computer graphics self-portrait painted by the flow of one's inner river of light. A person sits before a video camera. Her body is connected to a biofeedback sensor. She watches a real-time naturalistic image of herself on the video monitor. Information about her internal mind/body processes is digitized and conveyed to the central processing unit of the computer system. The video image is modified by a specially designed software package. It can be modified by changing color or size, by stretching, elongating, extending, rotating, replicating, superimposition or by other computer graphics effects. For example, the participant sees herself turn green and is shocked by the sight. The shock, in turn, changes the biofeedback information causing the computer to modify her self-portrait again. Her green face now becomes elongated. Changes in body processes affect changes in the video image. The perceived video image, in turn, stimulates the mind/body changes, and so on in a continuous feedback loop like the unending flow of a Torah scroll.[63]

This biofeedback artwork points to new directions for dialogic art of the future that will explore the dynamic interplay between carbon-based biosystems and silicon-based cybersystems. Living organisms are biosystems made up of numerous types of molecules built from versatile carbon atoms surrounded by a mass of water. The core of cybersystems are circuits etched on chips made from silicon atoms that are akin to carbon atoms. Carbon atoms with a 2-4 electron configuration sit in the same column in the Period Table of Elements as silicon atoms with 2-8-4 configuration. Having four electronics on their outer shells, both carbon and silicon atoms are amphoteric elements that can act as metals or non-metals. Elements with atoms of less than four electrons on the outer shell act as metals while those with more than four act as non-metals. Carbon atoms possess such amphoteric versatility that they form more compounds than of all the other 91 elements combined. Feedback loops between carbon-based plants, animals, and human consciousness and silicon-based artificial intelligence will provide the raw material for artists' creative endeavors.

In his paper, "Edge-Life: Technoetic Structures and Moist Media,"[64] Roy Ascott proposes that just as the development of interactive media replaced the cult of the object d'art with process-based art forms in the 20th century, future art forms will be dialogues between dry pixels and wet molecules. Art of the future will explore "moistmedia," the interspace between the dry world of virtuality and the wet world of biology in which spiritually numinous art forms will emerge. These new art forms will be conspicuously different from the art of earlier eras.

Notes

1. Arthur C. Danto, *Beyond the Brillo Box: The Visual Arts in Post-Historical Perspective* (New York: Farrar, Straus, Giroux, 1992), p. 5.
2. Wassily Kandinsky, *Kandinsky: Complete Writings on Art*, Kenneth Lindsay and Peter Vergo (ed.) (Cambridge, MA: Da Capo Press, 1994), p. 773.
3. *Babylonian Talmud: Avodah Zarah*, 41b.
4. Ibid., 44b.
5. Mel Alexenberg, "A Semiotic Taxonomy of Contemporary Art Forms," *Studies in Art Education*, 17: 3, 1976, pp. 7–12.
6. Charles S. Peirce, *Collected Papers* (Cambridge, MA: Harvard University Press, 1960).
7. Mel Alexenberg, "Semiotic Redefinition of Art in a Digital Age," in Debbie Smith-Shank (ed.), *Semiotics and Visual Culture: Sights, Signs, and Significance* (Reston, VA: National Art Education Association, 2004).
8. Lev Manovich, *The Language of New Media* (Cambridge, MA: MIT Press, 2001), p. 201, 202.
9. Ernest Namenyi, *The Essence of Jewish Art*, trans. Edouard Roditi (New York and London: Thomas Yoseloff, 1960).
10. JoAnn Wypijewski (ed.), *Painting by Numbers: Komar and Melamid's Scientific Guide to Art* (Berkeley: University of California Press, 1999).
11. *Breishit Raba*, 38:13.
12. Anthony Julius, *Idolizing Pictures: Idolatry, Iconoclasm and Jewish Art* (New York: Thames & Hudson, 2000), p. 71.
13. Svetlana Alpers, *Rembrandt's Enterprise: The Studio and the Market* (Chicago: University of Chicago Press, 1988), p. 105, 110.
14. Kynaston McShine (ed.), *Andy Warhol: A Retrospective* (New York: The Museum of Modern Art, 1989), pp. 160–167.
15. Lucy R. Lippard, *Six Years: The Dematerialization of the Art Object from 1966 to 1972* (New York: Praeger Publishers,1973), p. 122.
16. Charles L. Mee, *Rembrandt's Portrait: A Biography* (New York: Simon and Schuster, 1988), p. 238.
17. Immanuel Kant, *Critique of Judgment*, trans. J. H. Bernard (London: MacMillan, 1931), p. 55.
18. Edward Bullough, "Psychical Distance as a Factor in Art and an Aesthetic Principle," in Eliseo Vivas and Murray Krieger (eds.), *The Problems of Aesthetics* (New York: Holt, Rinehart and Winston, 1966), pp. 406-411.
19. Suzi Gablik, *The Reenchantment of Art* (New York: Thames and Hudson, 1991), p. 5.
20. *Talmud Bavli*: *Baba Batra*, 14b. Translated and adapted by the author.

21. Walter Benjamin, "The Work of Art in the Age of Mechanical Reproduction," in M. Jennings, M. Bullock, H. Eiland, G. Smith (eds.), *Walter Benjamin: Selected Writings, Vol. 3, 1935–1938* (Cambridge, MA: Belknap Press/Harvard University Press, 2002).
22. Carter Radcliff, *Out of the Box: The Reinvention of Art, 1965–75* (New York: Allworth Press and School of Visual Arts, 2000), p. 240.
23. Alexander Alberro and Patricia Norvell (eds.), *Recording Conceptual Art* (Berkeley: University of California Press, 2001), p. 17.
24. Adin Steinsaltz, "World, Angels, and Men," *The Strife and the Spirit* (Northvale, NJ: Jason Aronson, 1988), pp. 42–53.
25. John Naisbitt, *High Tech High Touch: Technology and our Accelerated Search for Meaning* (London: Nicholas Bealey, 2001).
26. Miriam Benjamin Alexenberg, *Clayscapes*, MFA thesis, School of Art and Design, Pratt Institute, 1987.
27. Paulus Berensohn, letter to the author.
28. Allan Kaprow, *Essays on the Blurring of Art and Life*, Jeff Kelley (ed.) (Berkeley: University of California Press, 1993), p. 223.
29. Wassily Kandinsky, *Concerning the Spiritual in Art*, trans. M. T. H. Sadler (New York: Dover, 1977), p. 52.
30. Ibid., p. 47.
31. Wassily Kandinsky, *Point and Line to Plane* (New York: Dover, 1979).
32. Anna C. Chave, *Mark Rothko: Subjects in Abstraction* (New Haven: Yale University Press, 1989), p. 1.
33. Don Judd, "Specific Objects," in J. R. Mellow (ed.), *Arts Yearbook: Contemporary Sculpture* (New York: Art Digest, 1965).
34. Barbara C. Matilsky, *Fragile Ecologies: Contemporary Artists' Interpretations and Solutions* (New York: Rizzoli in association with the Queens Museum of Art, 1992), p. 51.
35. Kaprow, *Essays on the Blurring of Art and Life*, p. 232.
36. Ibid., p. 221.
37. Ibid., p. 222.
38. Julian Jaynes, *The Origin of Consciousness in the Breakdown of the Bicameral Mind* (Boston: Houghton Mifflin, 1976).
39. Nosson Scherman, *The Complete Art Scroll Siddur* (New York: Mesorah, 1987), p. 2.
40. Tom Zeller, "Seen My Sock Drawer Lately? Check Out My Web Site," *The New York Times*, Weekly Review, August 19, 2001, p. 8.
41. Christo and Jeanne Claude, http://christojeanneclaude.net/si.shtml, Accessed October 7, 2010.
42. Lawrence Halprin and J. Burns, *Taking Part: A Workshop Approach to Collective Creativity* (Cambridge, MA: The MIT Press, 1974), p. 89.
43. Mierle Laderman Ukeles, "Maintenance Art Manifesto: Proposal for an Exhibition, 'CARE,'" in Alexander Alberro and Blake Stimson (eds.), *Conceptual Art: A Critical Anthology*, (Cambridge, MA: The MIT Press, 2000), p. 124.
44. Allan Kaprow, "A Statement," in Michael Kirby, *Happenings* (New York: E. P. Dutton, 1966), p. 46.
45. Ibid.
46. Ibid., p. 49.
47. Matilsky, *Fragile Ecologies*, p. 72.
48. Michaela Hajkova, *Mel Alexenberg: Cyberangels/Aesthetic Peace Plan for the Middle East* (Prague: Robert Guttmann Gallery of the Jewish Museum in Prague, 2004), p. 12.

49. Elisabeth Siddiqui, "Islamic Art," http://www.colostate.edu/orgs/MSA/find_more/ islart.html, Accessed October 7, 2010.
50. Abdul Hadi Palazzi, "What the Qur'an Really Says," http://www.templemount.org/quranland.html. Accessed October 7, 2010.
51. Muhammad Thompson and Nasima Begum, *Islamic Textile Art: Anomalies in Kilims*, Salon du Tapis d'Orient, http://www.turkotek.com/salon_00101/salon.html, Accessed October 7, 2010
52. Interview with Paul A. Weiss in M. Alexenberg, *Aesthetic Experience in Creative Process* (Ramat Gan, Israel: Bar-Ilan University Press, 1981), p. 197–202.
53. Tashibiah Sayyed, "Israel – A State of Mind," http://www.arabsforisrael.com, Accessed October 7, 2010.
54. Maurice S. Friedman, *Martin Buber: The Life of Dialogue* (New York: Harper and Row, 1960), p. 85.
55. Linda M. Montano, "Mierle Laderman Ukeles," *Performance Artists Talking in the Eighties* (Berkeley: University of California Press, 2000), pp. 454–459.
56. Laszlo Moholy-Nagy, *Vision in Motion* (Chicago: Paul Theobald, 1965).
57. Yaacov Agam and Bernard Mandelbaum, *Art and Judaism* (New York: BLD Limited, 1981), p. 22, 23.
58. Eduardo Kac, "Essay Concerning Human Understanding," http://www.ekac.org/Essay.html.
59. Eduardo Kac, e-mail communication to author, September 7, 2001.
60. Radcliff, *Out of the Box: The Reinvention of Art, 1965*–1975, p. 191.
61. Manovich, *The Language of New Media*, p. 66, 67.
62. Laurent Mignonneau and Christa Sommerer, "Designing Interfaces for Interactive Artworks," IEEE KES 2000 Knowledge Based Engineering Systems Conference Proceedings (Brighton, UK: University of Brighton, 2000), pp. 80–84.
63. Alexenberg, *LightsOROT*, p. 55, 56.
64. Roy Ascott, *Art, Technology, Consciousness: mind@large* (Bristol, UK: Intellect Books, 2000), pp. 2–6.

Chapter 3

Morphological Perspectives: *Space-Time Structures of Visual Culture*

How can we characterize the emergent field "visual culture"? To begin with, we must insist that this encompasses a great deal more than the study of images, of even the most open-ended and cross-disciplinary study of image. Visual culture opens up an entire world of intertextuality in which images, sounds and special delineations are read on to and through one another, leading ever-accruing layers of meanings and of subjective responses.[1]

Irit Rogoff

Whereas the aesthetic perspective oriented us to the making of objects, the ecological perspective connects art to its integrative role in the larger whole and the web of relationships in which art exists. A new emphasis falls on community and environment [...] The ecological perspective does not replace the aesthetic, but gives a deeper account of what art is doing, reformulating its meaning and purpose beyond the gallery system, in order to redress the lack of concern, within the aesthetic model, for issues of context or social responsibility.[2]

Suzi Gablik

Structures of space and time and the relationships between the senses in experiencing these structures form the basis of morphological analysis of visual culture and its art forms. Each culture is an expression of a structure of consciousness that gives rise to a particular perspective. The mythic, mental-rational, and integral structures of consciousness of civilizations proposed by Gebser[3] give rise to mythological, logical, and ecological perspectives. The hermeneutical method of critical analysis used by Gebser and Mickunas[4] to explore the morphologies of different cultures as experienced by those who live in them is a valuable methodology for understanding structures of consciousness and the perspectives that arise from them. The structure of consciousness that characterizes both Judaism and postmodernism is an integral structure of consciousness that gives rise to ecological perspective. Understanding the morphologies of mythological and logic perspectives makes it possible to understand the origins and evolution of ecological perspective, its modern and postmodern art forms, and its relationships to Jewish structures of consciousness.

A mythic structure of consciousness with its mythological perspective characterizes most hunting tribes, traditional agrarian societies, and pre-industrial cultures, both past and present. Two exceptions to this generalization about the mythological perspective of traditional cultures in the ancient world are the Greeks and the Hebrews. The Greeks developed rational consciousness with its logical perspective while the Hebrews developed integral consciousness with its corresponding ecological perspective. Ecological perspective, an anomaly among traditional cultures, was preserved for two millennia embedded in the culture of the Jewish people dispersed among civilizations with mythological and logical perspectives.

Artists shape worldview by their perspective inventions. Renaissance artists renewed the Greek logical perspective by visually representing three-dimensional space from a single point of view and time as a cross-section of a one-way linear path. Most people in the industrialized world continue to see the world through the eyes of these Renaissance artists. Before Renaissance perspective spread from Italy, medieval art throughout Europe expressed a mythological perspective shared by the art of tribal and traditional cultures worldwide. Most third world people continue to see their world through a mythological perspective, an auditory experience of space as two-dimensional and time as cyclical.

Artists today are once again reshaping humanity's worldview by inventing art of ecological perspective. Ecological perspective leads to seeing the lively interrelationships between parts of a whole that is more than the sum of its parts. Space and time are unified in a four-dimensional world of events experienced through movement and interaction expressed in art through lively narratives. Senses are integrated to gain holistic perception of the structure of a dynamic diaphanous world. This perspective has not yet permeated the thoughts and actions of most people who continue to see the world through the lenses of logical and mythological perspectives. The organic synthesis, interdisciplinary integration, and global reach of ecological perspective can think the world together in contrast with the cold analysis and the fragmentation of knowledge into discrete disciplines of logical perspective that think the world apart. Gablik proposes that there is a need for new forms emphasizing our essential interconnectedness rather than our separateness, forms evoking the feeling of belonging to a larger whole rather than expressing the isolated, alienated self.[5]

Latent structures and transformative processes

Gebser proposes five distinct structures of consciousness, each having a predominant mode of experiencing: archaic, magic-vital, mythic, mental-rational and integral. The mythic structure gives rise to mythological perspective, the mental-rational structure to logical perspective, and the integral structure to ecological perspective. New structures of consciousness do not surpass and replace earlier structures that are retained as latent structures.

Psychologist Julian Jaynes in *Origins of Consciousness in the Breakdown of the Bicameral Mind*[6] traces the emergence of biblical consciousness as a rejection of archaic and magic-vital structures. The Bible identifies acts that are expressions of these two structures and explicitly forbids them: "You shall not learn to act according to these abominations [...] anyone involved in these revolting practices is repulsive to God." (*Deuteronomy 18:9–12*). The archaic structure is akin to dreamless sleep in which the human is completely submerged in and coextensive with the world. According to Jaynes, it is a preconscious state when the human mind is unaware of its awareness. It forms the basis of idolatry in which the large eyes of Middle Eastern idols trigger one hemisphere of the brain to speak to the other hemisphere making the idolater believe that he is hearing a divine voice commanding him to act and instructing him on how to act. The archaic structure is a preconscious structure of zero dimensions and the magic-vital structure is a one-dimensional pre-perspectival structure.

> The magic-vital awareness is one of identity. Every event is vitally connected to and can be transformed into every other event; one can become the other. In vital awareness, the human has no specific egological identity or psychological self (image); rather, the human is identical with the powers that it embodies. Thus, the hunter who performs the hunted animal's movements in dance or wears the animal's skin consists of the very power of the animal. The hunter does not symbolize the animal as if s/he has a permanent identity separate from it and was merely *enacting* the animal.[7]

Gebser's concept of "ever-present origin" posits that the archaic and magic-vital structures are not destroyed in later structures of consciousness but are retained in latent forms with the potential of re-emerging. Therefore, Judaism created *halakhah*, an elaborate set of practical ways to override, redirect, and suppress these latent tendencies. Halakhic perspective is an ecological perspective transforming mythological and logical perspectives into an integral structure of consciousness. This millennia-old transformation reoccurring in postmodernism is transformative in the sense that logical perspective is not overthrown, but is transformed together with mythological perspective into a new structure.

I use the word "transformation" as it is used in art by my former Columbia University colleague Justin Schor in *Toward the Transformation of Art*,[8] and in science by George Lock Land in *Grow or Die: The Unifying Principle of Transformation*.[9] A transformative process reorganizes component structures into a new compound structure with characteristics that cannot be predicted from information presented by the components taken separately. Logical and mythological perspectives are transformed into a new compound ecological perspective in a process analogous to how photosynthesis transforms carbon dioxide and water into sugar. Sugar is a new compound that exhibits physical and chemical properties quite different from those of the two component compounds. However, the carbon, hydrogen, and oxygen atoms in sugar do not lose the identities they had in the component compounds before transformation. A compound structure is not merely a mixture, but a structure existing because of the interactive bonds between its component parts. A mixture of carbon dioxide and water is seltzer, not sugar.

Another way to look at the transformative relationship between components is to consider Pablo Picasso's assembled sculpture of a bull's head. Picasso took a bicycle seat with its narrow front pointed downward and connected handlebars to it so that they look like the horns of a bull emerging from the bicycle-seat head. The bicycle seat and the handlebars are transformed into a bull's head – a new compound structure formed by our perception of relationships between its components. Certainly, a bicycle seat and handlebars taken separately give no clue to the meaning of their relationship as a bull's head. Although the transformative process unifies, it does not homogenize. Multiple meanings are simultaneously available. The presentation of the bicycle seat and handlebars as themselves is given at the same time as the representation of a bull's head. Picasso's *Bull's Head* can be seen as a metaphor for the ecological paradigm. Aspects of logical and mythological structures retain their identities as they merge to form a new ecological paradigm.

MIT professor Thomas Kuhn, in *The Structure of Scientific Revolutions*,[10] proposes that old paradigms are often strengthened at the time when the new paradigm is emerging. Mythological and logical people who fear the new, lack flexibility, and have limited understanding, cannot shift perspective and, therefore, reject the ecological paradigm. Since people who experience the world though mythological or logical perspective cannot conceive of an ecological perspective, the Jews were viewed as aliens to be feared, hated and persecuted. This inability to see from an alternative perspective is described in Edwin Abbot's classic novel *Flatland*.[11] Flatland is a world that only has two dimensions inhabited by polygons that can only see each other as lines from a perspective within a flat plane. When a Flatlandian conceptualizes a three-dimensional world from which one could look down at the plane and see the shapes of the polygons, he is condemned and jailed.

Just as a three-dimensional world is viewed as a dangerous idea by Flatlandians, a four-dimensional world of multiple viewpoints integrating time and space in ecological perspective is incomprehensible and feared by those trapped in the three-dimensional logical perspective of European culture. The four-dimensionality and beyond that characterizes Jewish consciousness and Einsteinian physics, arose in modern art and flourishes in postmodern art. Aspiring artist Hitler mounted his *Degenerate Art* exhibition in 1937 in Munich as a condemnation of the emerging ecological perspective in modern art calling it "Jewish Art" even though gentiles had made most of the artworks in the show.[12] Einstein's theories were labeled "Jewish Physics" and banned. European civilization's fear of those with alternative perspectives generated centuries of anti-Semitic hatred that led to the Holocaust. In Nazi Europe, latent magic-vital and mythic structures of consciousness bubbled to the surface to disfigure the predominant mental-rational structure and create horrific distortions of it. The German culture that represented the height of a rational structure of consciousness and logical perspective had the industrial organization to assure that genocide was well planned and efficiently executed. A "final solution" was devised to rid the world of the Jews who brought ecological consciousness to humanity.

Alternative perspectives

The morphological difference between mythological and logical perspectives struck me in 1969 when I moved with my family from New York to a cottage set in an orange grove in Israel. Our neighbors were Yemenite Jews who had ascended to Israel from the tip of the Arabian Peninsula a decade earlier. These Jews had lived for 2,000 years in a style of life that changed little from biblical times. Although the ecological structure of Jewish consciousness remained embedded as the deep structure of Jewish culture in Islamic lands, centuries dwelling among mythological people had influenced their perspective in surface ways, just as European Jews had their integral worldview distorted by the overpowering logical perspective of a Western culture shaped by Hellenism.

Having returned to their biblical homeland, they continued to bake flat round bread in a wood-burning, underground oven in their back yard as they had done in Yemen. Their daughter, Zahava, came to welcome us with two large pita-like breads, one in each hand. They were still warm and surrounded by the welcoming aroma of fresh-baked bread. Compare these whole, two-dimensional, circular breads that Zahava brought us to the supermarket bread of the industrialized world. Supermarket bread is a three-dimensional, rectilinear, cold, white loaf fragmented into slices and kept at a distance from the consumer by a sealed plastic wrapper that cuts off olfactory and tactile contact. The space-time morphologies of mythological cultures are two-dimensional, circular, and cyclical as symbolized by pita-like breads. Supermarket white bread symbolizes the three-dimensional rectilinear space and linear time of logical culture.

The two spiral loaves used for celebrating the Jewish Sabbath symbolize the space-time morphologies of the integral structure of consciousness that gives rise to ecological perspective. The spiral form is an open-ended growth form unlike the closed circle of the pita or rectangular slices of supermarket white bread framed by its crust. In Jewish consciousness, the circle symbolizes idolatry and the rectangle slavery. Both are closed forms. The Hebrew word for "brickyard," where the Israelites had been enslaved, is also the word for "rectangle." The Hebrew word for "calf" in the story of the Golden Calf can also be read as "circle;" both are spelled with the same letters. Only weeks away from slavery, the Israelites certainly knew Ra, the supreme deity of Egypt, the sun god represented by a golden circle that we can see in ancient Egyptian art.

The Sabbath bread is either shaped like a spiral snail or braided like the DNA helix. It represents living systems open to growth rather than the closed circle of pita or the rectangular slices of supermarket bread. The bread baked in haste on leaving Egypt is commemorated on Passover by eating flat, round, hand-made mythological *matzah* or rectangular machine-made logical *matzah* called "bread of affliction." Leavened bread called *h̲alah* symbolizes the showbread baked in freedom for the Tabernacle ritual in the Sinai desert (*Leviticus 24:5*). Paralleling the postmodern concept of double-coding, *matzah* can also be viewed as bread of humility as opposed to the puffed-up pretension of *h̲alah*. The multiple perspectives of Jewish consciousness validate simultaneous seeing from contrary viewpoints.

Tuna sandwiches on supermarket white bread or hummus scooped up with pita are usually eaten without seeking, acknowledging, or honoring the complex web of interrelationships linked to the spiritual significance of our lunch. However, with *h̲alot* (plural for *h̲alah*) we cook up a postmodern meal that links the act of eating with celebrating the connections of bread to its spiritual place in a global ecosystem, to the Creator of the universe, to freedom from slavery and idolatry, to open-ended growth systems, to rewards for righteousness and compassion, and to the partnership between humanity and God in the continuing story of creation.

On the Sabbath, two whole, unsliced *halot* rest on the table hidden beneath a beautifully embroidered covering. We place two loaves side by side to recall the double portion of manna that rained down every Friday so that the Israelites could celebrate the Sabbath day without having to work to collect their food (*Exodus 16*:5). The two *halot* are covered to hide them from insult by not allowing them to see us saying the blessing over wine before we say the blessing over them. The bread and wine are illuminated by flaming wicks in olive oil or candles to acknowledge the divine blessing promised for living a life of righteousness and loving kindness. "I shall provide rain for your Land in its proper time, the early and the late rains, that you may gather in your grain, your wine, and your oil" (*Deuteronomy 11:14*). Before touching the *halot*, everyone washes hands with a two-handled cup in a ritual fashion reciting a blessing on the uplifting of our hands. Then, the *halot* are uncovered and held together while saying a blessing linking these breads to the Creator of the universe who brings forth bread from the earth. Only after acknowledging the spiritual place of the bread in our planet's ecosystem, can the bread be broken for everyone to have a piece to eat as the beginning of the Sabbath meal. Pinchas Peli offers an additional reason for celebrating with wine and *halot* that relates to the creative work of the artist:

> Wine and bread clearly demonstrate the partnership of humanity and God in the act of creation. Neither wine not bread grow on trees. It is God who created the vine and we who extract wine from it; God who makes corn or wheat and we who make bread of them. Does not the last word in the story of creation which introduces us to the Sabbath – *la-asot* – state that God created the universe for humanity to continue making it? What better way is there to celebrate creation than to partake of products which are the combined result of the creation of God and the handiwork of human beings? [13]

While promoting and celebrating integral associations, multiple meanings, and deep significance of wine and *halah*, wine remains a tasty drink to relax us at the Sabbath meal and *halah* remains bread not unlike a supermarket loaf or pita. Spiritual and material aspects of wine and *halah* are experienced simultaneously. They never attain magical or mystical dimensions. Spirituality emerges from the quality of our perception of the material world. Judaism and postmodernism share the goals of exploring multiple viewpoints, reading alternative codes, deconstructing and reconstructing meanings, and creating new narratives.

Mythological perspective

In a world of mythological perspective, people sit together within a two-dimensional, circular, and cyclical world. They experience this world from hearing the retelling of the oral tradition – their communal mythology as handed down to them by word of mouth. They sit as a community surrounded by a sphere of sound and often wrapped with a dwelling made from a single two-dimensional plane formed into a cylinder, cone or dome.

The auditory experience of space is encircling, involving, and soft-edged. Time is felt as cyclical and pulsating. The nature of the auditory experience is derived from the physics of sound. Sound generated in air produces spherical waves that surround the point of origin and engulf anyone within its sphere. A cross-section of a sound-sphere would appear like the concentric circles that surround a pebble when it is tossed in a pond. When people sit within a sound-sphere of pulsating air, they cannot help hearing the message. They feel that sound surrounds them and involves them. They can neither turn away from it nor close their ears to it; ears have no anatomical analogues to eyelids. Unlike the visual world where light sources can be accurately pinpointed, the auditory world is soft and fuzzy at both its core and edges.

"The 'wrap-around' aspect of auditory space," anthropologist Edmund Carpenter relates, "is shown by the manner in which an Eskimo constructs an igloo. Surrounded by space in all its acoustic non-direction, he does not mould his igloo from outside looking in, but from inside working out." He is in the center of a surface of snow that he builds around himself with a series of concentric circles tapering upwards conically. In summer, he wraps skins around himself to construct a tent dwelling. Both igloo and tent surround the Eskimo with a single, two-dimensional, circular place. He does not know the vertical walls and horizontal ceilings that meet at ninety-degree angles in the carpentered world of logical perspective.

In his extensive studies of Eskimo culture, Carpenter claims that he has found no examples of Eskimos describing space primarily in visual terms. Since they do not regard space as being static or time as linear, they have no units of special measurements or uniform division of time. The Eskimo carver holds a piece of ivory in his hand and asks it to tell him what hides within. He carves receptively, humming or chanting as he works, until the hidden form emerges. He does not create nor make an ivory seal, for example. He releases the seal hiding in the ivory by listening to what the ivory has to say to him as he carves.

> The carver is indifferent to the demands of the optical eye; he lets each piece fill its own space, create its own world, without reference to background or anything external to it. Each carving lives in special independence. Size and shape, proportion and selection, these are set by the object itself, not forced from without. Like sound, each carving creates its own space, its own identity; it imposes its own assumptions.[14]

Form exists, the Eskimo poet says "on the threshold of my tongue." Unlike logical people to whom "seeing is believing," Eskimos trust their ears rather than their eyes. The ocularly visible apparition is less important than the auditory truth of oral tradition. "A carving, like a song, is not a thing," Carpenter writes, "it is an action." Just as one sings when he feels a song within him, so the Eskimo releases a carving when he senses a form emerging from the ivory. When spring comes and the igloos melt, carvings can be found among the wastes littering the old habitation sites. These carvings were not deliberately thrown away; they were just lost. In contrast to logical man's valuation of the art product, the Eskimo is indifferent to the product; he values the artistic act.

The Eskimo carver makes no effort to show a favored point of view in his carvings. He is neither interested in developing a personal style nor in presenting a single viewing position. He attempts to let the ivory speak through him in the grammar of timeless tradition. Many of his carvings, generally of sea mammals, lack a base and will not stand up. The carving is released from the ivory to live in his hand and in the hands of others as it is moved and spoken to and about. Walrus tusks are usually carved into aggregates of connected but unrelated figures facing in different directions. The Eskimo carved the figures, each oriented in another direction, without moving the tusk.

Driving in South Africa, I saw children at the side of the road holding out wooden giraffes carved from long branches with markings painted with berry juice. I stopped to buy one. Each giraffe had its four long, skinny legs emerging from a carved base. I tried in vain to find one that would stand up. When I asked the children why they had carved bases if none of the giraffes would stand, they explained that they made the bases only to satisfy the aesthetic preference of white tourists. Bases are for logical people. Mythological people have no use for them. In *Culture and Thought*,[15] Cole and Scribner describe how tribal African children draw figures all over a page, turning it around and around so that all four edges become baselines.

Mythological people do not see depth in pictures that logical people see as being three-dimensional. A number of studies based upon Hudson's test for identifying 2D and 3D perceivers have show that mythological people in many parts of Africa and from a variety of tribal and linguistic groups are unable to see the third dimension in pictures. Hudson's test employs a set of seven picture cards, each showing a drawing that includes an elephant, an antelope, and a man pointing a spear. In each picture, the three figures are placed in a different landscape. Each landscape provides at least two of the following depth cues: (1) Familiar size. The elephant, which is known to be larger than the antelope, is drawn smaller than it. (2) Overlap. One object is drawn obscuring a portion of another object. (3) Linear perspective. Two lines, which are known to be parallel, are drawn converging. In each picture, the spear is aligned so that it points at both the elephant and the antelope. A subject is shown one picture at a time and is questioned about the relationships between the figures. He is asked: "At which animal is the man aiming the spear?" "What is closer to the man?" One of the pictures shows a small elephant on a hill in the center background with a larger antelope in the right foreground and a man holding up his spear in the left foreground. The tribal African explains that the man is aiming his spear at the elephant that is closer to him, being in the center of the picture rather than at the far right. He perceives all three figures as being on the same plane. To logical people, the man is aiming his spear at the antelope facing the man on the same plane, not at the small elephant standing on a distant hilltop.[16]

That logical people actually see depth in a picture that mythological people cannot see as three-dimensional was demonstrated by Jan Deregowski.[17] He displayed the Hudson picture on layers of transparencies in a box and asked each subject to move a light beam to coincide with the depth of each of the three figures. Europeans set the light further away when setting it over the elephant than when setting it over the other two figures, showing they were not merely interpreting conventional artistic cues but actually saw depth in the picture. The tribal Africans set the light to indicate that they saw all three figures on the same plane.

Mythological people in the preliterate cultures of medieval Europe did not look at pictures to see space defined. They listened to church paintings tell them the story of Christianity. Medieval ecclesiastical paintings were didactic art aimed at educating the illiterate populace in the rudiments of Christine doctrine. Arnold Hauser writes in *The Social History of Art*:[18]

> In the Middle Ages art would be superfluous if everyone could read and follow an abstract chain of reasoning; art was originally looked upon just as a concession to the ignorant masses [...] It was certainly not allowed to be a "mere pleasure of the eye," as St. Nilus put it. Its didactic character is the most typical feature of Christian art.

The two-dimensional paintings, lacking any illusion of depth, in the churches of Western Europe were paralleled by the icons in the churches of the East in their educational function. Art historian Michelis reaches the same conclusion from his study of the religious art of the Eastern churches: "Byzantine painting was not only a religious but a didactic art. Its object was by its icons to teach the Christian Orthodox religion even to the illiterate. In the words of Basil the Great; 'What the verbal account presents to the ear the silent picture reveals to imitation.'"[19] Unlike Renaissance paintings' illusion of three-dimensional space for disinterested contemplation, two-dimensional medieval paintings related the traditional mythology of the church for auditory involvement.

Logical perspective

Logical perspective has one standing alone outside of a world of objects that are set in particular niches in three-dimensional space. From a single vantage point, one looks at a fragment of the world arrested at a point on the shaft of time's arrow. One gets to know the world visually, from rays of light traveling to one's eyes in straight lines from definite points in Euclidean space.

Although this perspective began to germinate in ancient Greece, it lay dormant through the Middle Ages to be reborn in Renaissance Europe at the hands of the painter Giotto (1267–1337). He broke the two-dimensional picture plane with his brush by painting an illusion of a three-dimensional space that appears to recede into the picture. In his painting *Madonna Enthroned*, Giotto placed the Madonna on a stage in the center of the panel with figures on the outside, on both sides of the stage. The front of the stage is free of painted figures. It is a place for us to stand. We are invited to enter the painting that seems to be a continuation of our real world. Then we are denied entry. We find ourselves trapped in the single spot demanded by the painter for us to see his uniperspective world from a predetermined distance.

Logical perspective developed in the paintings of Masaccio (1401–1428). In his fresco *Trinity* in Florence, Masaccio provided enough data to measure the depth of the painted interior chamber. All the lines in the painting perpendicular to the picture plane converge at a single point at the base of the cross. To see the fresco properly, we must face this point that is at eye level some five feet above the church floor. Massaccio demanded our submission to his singular point of view. We are asked to separate ourselves from the world we see as real, to be passive spectators rather than active participants, and to see the objects that comprise the world within a rigid orthogonal framework. Visual perception of space is primary.

Visual perception isolated from our other senses is a biologically distancing experience. Only the visual sense requires that we keep objects away from us in order to focus our eyes on them. A child can hear her father's heartbeat with her ear against his chest. She can smell flower petals touching her nose, taste sweets touching her tongue, and feel her stuffed bear by caressing it with her hands. The auditory, olfactory, taste, and tactile senses can put us in direct and intimate contact with the physical world. The visual experience, disconnected from the other senses, is alienating by its very nature, demanding that we keep our distance in order to see. Moreover, vision is the only sense modality with an anatomical mechanism that can shut out the world. We have eyelids.

In logical perspective painting, objects are shown to cast exact shadows to indicate a particular moment in time. This fixes the fragment of the world depicted in the painting at a single point on a one-dimensional and unidirectional temporal track. Time in logical perspective is seen as a linear track along which objects move at uniform speed from past through present into the future. It is an irreversible continuum that is segmented into equidistant units. Objects moving along this track, Louis Rapkine argues, increase in entropy, in disorder, and in probability of disintegration.[20] Since logical people see themselves as objects being dragged along the one-way track to ultimate death, they become obsessed with stopping the flow of time. They use a paintbrush as an axe to chop out a cross-section of the track as an immortalized moment. They paint the cross-section with durable materials that are "permanent," so that it will have eternal existence and value. A dynamic flow is stopped and preserved as a static image. They believe they can conquer nature and force its flux and vigor into a rectilinear frame. The frame not only stops the flow of time by containing it, but it also separates the immortalized moment from living experience in the real world. The frozen moment is deified by surrounding it with the halo of an ornate golden frame. The individual artist gains immortality by affixing a unique signature to each painting.

An alternative response of logical people to the continual flow of time is to glorify the oncoming rush of the future. "The avant-garde is dead; long live the avant-garde!" This is the cry of the artist who deifies the section of the time-track that he is approaching rather than a cross-section of a part of the track past. Most logical people express their strongest affirmation of faith in progress, however, not in art but in science and its spin-off technology. In science, new conceptual models replace old ones; and these new conceptual models will be replaced in the future with still newer ones. In technology, the new model car is an improvement over the older model; and in the future, the newer model will be even better.

In order to better grasp the significance of logical perspective in art, it is instructive to examine its parallel in science. Logical perspective in both art and science share a common Hellenistic origin. The Athenians and Ionians developed what M. C. Goodall[21] calls the "First Science." The paradigm of the First Science employs a mental logic by the application of a logical system of inference. The theorem of Pythagoras was a logical advance along the line of progress from earlier axioms of Euclid. In the Renaissance, the mental logic gave rise to the empirical logic of a "Second Science." The basic methodology of the Second Science is to interrogate nature through a logical sequence of experiments. Questions put to nature in such a way that the answers will be "mechanistic rather than vitalistic and causalistic rather than teleological." The mechanistic view, according to biologist Paul Weisz,[22] holds that all physical and chemical phenomena in the universe can be accounted for with no other phenomena remaining. Therefore, the controlling agent of the material in the universe must reside within the material itself. Moreover, Weisz summarizes, the entire universe including life must consist of physical and chemical events only. These events occur in a linear sequence. The occurrence of an event is caused by a prior event. Causalism maintains that events are linked in a cause and effect chain in which end states are consequences. In contrast to this logical perspective, mythological perspective maintains teleology, which uses ultimate purpose or design as a means of explaining natural phenomena.

Logical artists and scientists stand outside of the box of nature attempting to look in, while mythological people sit inside the box listening to their elders telling traditional legends about nature. Logical scientists visually study a mechanistic and causalistic world looking in on it with an aloof air. Logical perspective in science reached its zenith at the close of the nineteenth century when the "billiard ball model" of the universe flourished. It was a time, historian of science A. D. Richie points out, "When every gentleman owned a billiard table and even those who were not gentlemen played billiards [...] It is the illusion of the visualist that he can observe completely objectively, like a spectator who is not a performer in the game."[23] To understand molecular behavior, scientists visualized molecules as billiard balls interacting on a billiard table. They did not deal with intentions, energies, and actions of the men holding the cues. Einstein transformed the logical game into an ecological one in the first decades of the 20th century while Picasso was creating Cubism and Kandinsky was proposing an abstract art of interrelating forms and colors.

Ecological perspective

The word "ecological" is appropriated from the biological sciences as the name for the perspective associated with an integral structure of consciousness. Like ecologists who look for patterns of interaction between animals and plants and their shared environment, ecological perspective facilitates our seeing dynamic patterns of interaction between space and time in our physical, mental, and spiritual environments. It honors multiple views of the whole and of how its parts interact to form an integral structure. Al Gore defines ecological perspective in his book *Earth in Balance: Ecology and the Human Spirit*:

> The ecological perspective begins with the view of the whole, an understanding of how the various parts of nature interact in patterns that tend towards balance and persist over time. But this perspective cannot treat the earth as something separate from human civilization; we are part of the whole too, and looking at it ultimately means looking at ourselves.[24]

Ecological perspective seeks to perceive, understand, and appreciate the complex interdependences of systems that include humanity with its technological creations and the totality of nature. Unlike mechanical technologies of the industrial age, electronic technologies of the digital age are less invasive and destructive to natural ecosystems. Artist Jurgen Claus coined the word "ecotechnology" to describe the interaction between technology and ecology.

> Ecotechnology means applications of instruments, materials, processes of technology in such a way that it truly harmonizes with nature, with the habitat of plant, animal, man, also with wider zones of our ecological home, indeed, with the entire globe and the cosmic space. A concrete example of ecotechnology is the use of light as a source of energy, photovoltage, which, like other technologies of our time, effects the transition from the mechanical principle (steam turbine) to the electronic principle.[25]

Ecological perspective sees through all realms of human culture creating an integral structure of consciousness that renders independent disciplines created by the logical mind diaphanous. This transparency invites us to look through disciplines to reveal underlying structures that unite them that are invisible to a mental-rational structure of consciousness. Ecological perspective renders boundaries between disciplines as permeable as cell membranes that facilitate the lively interchange of information. Art of integral consciousness and ecological perspective breaks open the frame to create a vital dialogue between multiple realms of discourse through active participation in a dynamic ecosystem of interrelationships flowing between real space and cyberspace.

To explore these ideas pedagogically, I developed a graduate course, *Morphodynamics: The Design of Natural Systems*, which I taught at Tel Aviv University and at Columbia University. I coined the word "morphodynamics" rather than use the word "morphology" that is too static. Morphodynamics is the study of dynamic processes that give rise to forms. At Tel Aviv University, I also developed a morphodynamics curriculum for Israeli schools, *From Science to Art,*[26] which developed student activities to bridge the gap between Jewish children with parents from logical European cultures and mythological Islamic cultures. Lozano and Mickunas describe my *From Science to Art* curriculum project in the book *Consciousness and Culture* as an exemplary model of integral pedagogy applicable to encounters between mythological and logical perspectives in a common educational context.[27] Morphodynamics activities weaving across the interface between science and art attempted to develop an integral consciousness and ecological perspective in Israeli youth of diverse cultural backgrounds. Through interplay between science and art activities, students explored the formation of spiral, branching, symmetrical, fractal, rhythmic, and other patterns in nature and culture. The curriculum facilitated developing an ecological perspective that related to the emerging postdigital era while encouraging their shared integral structure of Jewish consciousness to bubble up to the surface after being plastered over by centuries of living in European and Islamic lands.

The following description of a unit of study on "threshold phenomenon" in my Columbia University course is exemplary of the other units both in my university *Morphodynamics* courses and in the *From Science to Art* project. My Columbia course was an interdisciplinary course open to students from all the university departments. My classroom was a combination science laboratory and art studio. The science activities were designed as protoart experiences to provide stimulus and material for art-making.

We experience threshold phenomenon at the subatomic level when we walk on carpeting on a dry winter day and receive a shock as we touch the doorknob. We rub electrons off the carpet onto our feet. As negatively charged subatomic particles they repel each other, spreading out and coating our entire body. As our hand nears the doorknob, the electrons fly off our body all at once through a high voltage spark leaping the gap between our hand and the doorknob. The threshold phenomenon is an all-or-none action. The flash only leaps from our body to the doorknob at the moment the gap between them becomes small enough to be bridged. Then our body totally loses all its charge all at once. This same process is the principle behind the Van de Graaffe generator that can boost charged particles to an energy level of ten million electronvolts using a moving carpet-like belt and a metal dome on which to collect the subatomic particles. Scientists use this electrostatic generator to study nuclear forces.

In animals, a nerve cell fires when the stimulus reaches a high enough level to trigger a shift in electrical potential along the neuron. If the stimulus is too weak nothing happens. Only when the stimulus threshold is reached does the nerve cell fire. At the molecular level, a dripping faucet demonstrates the threshold phenomenon. The water builds up in the faucet until it is heavy enough to form a drop and fall. Then there is no water left in the faucet and the whole process repeats itself.

My students equipped with a video camera went down to the ferry terminal at the southern tip of Manhattan to document the threshold phenomenon at the level of living organisms. The Staten Island ferry had just left and the terminal building was empty. Gradually more and more people came into the building until it was packed. When the next ferry came and the doors were opened all the passengers rushed onto the boat completely emptying the terminal. Then more people started to come and the process began again. At a bus stop or gate in an airport terminal, we could document the same threshold phenomenon.

I showed my students photographs of all-or-none gathering at the organism level I had made in a turkey coop in Kibbutz Tirat Tzvi were I had worked as a farmer during the year before coming to Columbia. When I put out round red plastic dishes of feed in the morning, all the turkey chicks rushed to them at once. There were no chicks at all remaining in other parts of the coop.

At the level of human cognition, the students recognized how the earliest stage of the creative process exemplifies the threshold phenomenon. They had all experienced a new idea popping into consciousness fully formed as if out of nowhere. They learned more about this all-at-once flash of insight from the transcriptions of interviews of prominent artists and scientists in my book *Aesthetic Experience in Creative Process.*[28] The threshold phenomenon in the thought process of artists and scientists is the sudden emergence of a new idea in a prepared mind emptied of conscious striving. It is the Wisdom (*h̲okhmah*) stage of creativity in the Biblical narrative of the artist Betzalal building the Tabernacle in the Sinai desert. "I have filled him [Betzalel] with a divine spirit, with *Wisdom*, Understanding and Knowledge, and with talent for all types of craftsmanship" (*Exodus 30:3*). We will explore in depth in the Kabbalistic Perspective chapter how *h̲okhmah* in the world of creation requires the nullification of the ego to make oneself an empty vessel open to receiving a new insight. In the model of creative process developed by kabbalists in the Land of Israel in the 16th century, the *h̲okhmah* experience when a new insight suddenly enters consciousness is a spiritual exemplar of the threshold phenomenon.

Ecological perspective sees a world in which boundaries between disparate realms become diaphanous in contrast to logical perspective, which honors disciplinary boundaries that are opaque and well defined. The ecological perspective arising from an integral structure of consciousness seeks interrelationships between the generation of electrostatic energy, nerve physiology, a dripping faucet, boarding the Staten Island ferry, turkey chicks feeding, cognitive processes, and biblical secrets revealed through *kabbalah*.

Origins of ecological perspective in science

The origins of ecological perspective can be traced to Einstein's publication of his Special Theory of Relativity and the first paintings made by the Cubists in the first decade of the 20th century. The ecological perspective in modern art emerged and developed at the same time and pace as in science. Since it derives from an integral structure of consciousness that dissolves barriers between disciplines, it is meaningful to trace the transition from a logical to the ecological paradigm in both science and art. The paradigm shift from Newtonian to Einsteinian physics and from taxonomic biology to ecology and biochemical genetics parallels changes in the arts that led to postmodernism. The ecological zeitgeist revealed through art and science in our digital age provides the strongest evidence for the postmodern shift from Hellenistic to Hebraic consciousness.

In the Special Theory of Relativity, Einstein regards time as a fourth dimension linked to the three dimensions of space to form a unified space-time continuum. Rather than locating objects in 3D space and in linear time separately, he locates 4D events in an integral structure of space-time. The linear time series "past-present-future" breaks down when an event is observed simultaneously from different viewpoints. An event happening at the present for me, can be viewed as having happened in the past by a second observer and may still be in the future of a third observer. Conclusions about the same event depend upon the motion of the observers relative to each other. The logical perspective of seeing stable solid objects was softened by Einstein's concept that mass is a form of energy and that mass is a variable quantity increasing with an increase in its velocity. Even measuring instruments, considered to be invariable standards for logical people, change in relation to their motion; measuring rods shorten and clocks slow down when they move at high speeds approaching the speed of light.

Matter was further softened by Louis de Broglie's concept that electrons and other subatomic particles can simultaneously be waves. Although the distinction between particles and waves seem obvious in our visible world of everyday objects, in the minute world of invisible events the distinction becomes hazy. Logical perspective's single point of view gives way to a dual perspective in the question "Is it a particle or a wave?" goes unanswered. It is not experimentally possible to decide whether an electron is a particle or wave. The Uncertainty Principle proposes that it is impossible to determine simultaneously the position and velocity of an electron. The measurement of velocity becomes very inaccurate when we try to get the position more accurately, and when we try to measure velocity more accurately the position becomes uncertain. The indeterminateness is not the result of the impossibility of observing electrons without disturbing them. The information an electron can give about itself is limited. This limit gave rise to Max Plank's concept of the quantum and its development into an equation to calculate the probability of an electron being at a particular location in space-time. Determining the likelihood of an event taking place became the aim of quantum mechanics instead of attempting to pinpoint a definite position in time and in space.

In rethinking quantum mechanics since the time he was Einstein's colleague at Princeton, physicist David Bohm has developed a holographic model of the universe that extends the ecological paradigm in directions that will facilitate understanding the deep structure of Jewish consciousness as revealed by *kabbalah* in the next chapter. He proposes in *Wholeness and the Implicate Order*[29] that the entire cosmos is made out of a seamless holographic fabric in which subatomic particles and everything else in the universe are no more separate from one another than the components of the pattern in an oriental rug. Electrons are no more substantive than the shape a geyser takes when it gushes up out of a pool of water. Bohm calls the geyser that we see the *explicate* or unfolded order in contrast with the unseen deeper level of reality that shapes the geyser and moves it up out of the pool which is the *implicate* or enfolded order. He uses the hologram as a metaphor for the universe. To the naked eye, a holographic image on transparent film looks like concentric rings that form when a pebble is tossed in a pond. When a laser beam shines through the film, however, a three-dimensional image of shocking reality appears. Unlike a photographic negative, we can cut up a hologram and each of its pieces retains the entire image. All the information needed to project the image is contained in every part. Michael Talbot writes in his book *The Holographic Universe*:

> One of Bohm's most startling assertions is that tangible reality of our everyday lives is really a kind of illusion, like a holographic image. Underlying it is a deep order of existence, a vast and more primary level of reality that gives birth to all the objects and appearances of our physical world in much the same way that a piece of holographic film gives birth to a hologram.[30]

Stanford University neurophysiologist Karl Pribram, author of *Languages of the Brain*,[31] arrived at a holographic model of the brain in which memories are repeated and distributed throughout the brain as a way to explain the fact that people who had portions of their brains removed surgically did not suffer the loss of specific memories. Biologist Paul Pietsch provided experimental evidence supporting the holographic brain model although he had started out trying to prove Pribram wrong. He discovered that when he removed the brain of a salamander without killing it and then replaced it, the salamander resumed normal feeding behavior. When he replaced the brain upside down or switched the right and left hemispheres, the salamanders returned to being able to feed themselves. In over 700 operations, he sliced, flipped, shuffled, subtracted, and even minced the brain of his subjects and found that in all cases the salamanders could still feed normally.[32]

A whole living organism exemplifies the holographic concept since all its genetic information is repeated in all of its cells. Every cell contains the full set of DNA instructions for the embryonic development, growth, and functioning of every animal and plant. All instructions in all living things are written with the same four letters arranged in four words: AT, TA, CG, GC. Through biochemical genetics, we have come to understand that our links to all life forms runs deeper than our interrelationships in our shared environment. We are one family – all humanity, all animals, all plants – intimately linked on the inside genetically and on the outside ecologically. Ecological perspective renders nature diaphanous so that we can see through it to discover our inner and outer links to it.

Ecological people interact *with* nature, in contrast with logical people who act upon nature and mythological people who are acted upon by nature. They engage in dialogue with nature. Dikes in Holland are made with layers of mud and rocks and woven willow mats. When the fury of the North Atlantic strikes, these dikes absorb the force with the flexibility of willow branches by moving in tune with the waves. This ecological solution stands in contrast to the logical solution of most European port cities that have built sea walls of steel-reinforced concrete to stop the waves. Acting against nature, these rigid walls are eventually smashed apart and need to be rebuilt unlike the Dutch dikes that flexibly move with the natural rhythms. The mythological solution is to passively accept the edict of nature by neither building rigid walls nor flexible dikes. Following the mythological solution, one third of Holland would be under water.

Ecological perspective seeks to understand and appreciate the complex interdependences of natural systems and the laws that govern them so that we gain the knowledge to act in concert with nature. This idea is elegantly phrased in the last lines of Charles Darwin's *The Origin of the Species*:

> It is interesting to contemplate a tangled bank, clothed with many plants of many kinds, with birds singing in the bushes, with various insects flitting about, and with worms crawling through the damp earth, and to reflect that these elaborately constructed forms, so different from each other, and dependent upon each other in so complex a manner, have been produced by laws acting around us.[33]

Origins of ecological perspective in modern art

The ecological perspective, reaching its adolescence in the current epoch of electronic media and postmodern art, was born a century earlier in Picasso's art and Einstein's science. Picasso deliberately drew on mythological and logical perspectives in creating the beginnings of a new ecological perspective. This transformation is clearly evident in his 1907 painting *Les Demoiselles d'Avignon*. His intensive training in methods of European perspective painting coupled with his study of tribal sculpture from Africa and Oceania led him to create an artwork that demanded that we shift our way of seeing. In the painting, French demoiselles, some assuming Africa-mask faces, weave in and out of the picture space as hard-edged visual space dissolves into soft-edged auditory space. Experiencing African and Oceanic art invited Picasso to jump, Alice-in-Wonderland style, through a looking glass reflecting the logical world into a world of alternative space-time structures. Or as the art historian Janson puts it: "Picasso used primitive art as a battering ram against classical conceptions of beauty."[34] Although Picasso battered logical perspective, he linked this disintegrative action with a corresponding integrative action. He deconstructed the logical picture plane into fragments that became the raw materials for reconstructing it into an ecological space, much as he used fragments of a bicycle to construct a bull's head. He transformed static forms into a vibrant ecosystem of interrelationships.

Braque joined Picasso in presenting the illusion of rapid rotation that displays all sides at once. They painted multiple viewpoints to be perceived simultaneously as if the rapidly rotating forms had left their images behind. A multiperspective illusion was created by superimposition of changing views as if they were diaphanous. The interrelated forms are lifted out of the picture plane as our eyes scan the painted forms and see through them. Picasso and Braque made the picture plane leap further into the viewer's space by pasting overlapping layers of everyday materials on their canvases. The real-life stuff in these Cubist collages pushes the picture out into our real-life environment to grab us and pull us into it.

The juxtaposition of painted forms and found pieces of paper with words and images printed on them blurs boundaries between the real world and a painted illusion. The interactive space of Cubist collages weaves in and out of the picture plane and shifts between real and illusionary worlds.

Rosanna Vitale in her essay "The Significance of Aperspectival Art in Light of Gebser's Work" proposes that Cubism is "the most prolific illustrator of Gebser's concept of the diaphanous quality of the new consciousness that may be achieved through integration of time and space."[35] Gebser uses the term "aperspectival" to mean the simultaneous grasp of multiple perspectives. Making the Renaissance depth illusion of space recede into the picture while simultaneously inverting it and extending it outward creates the aperspetictival experience. Presentation of ordinary scraps as themselves coupled with the same scraps cut, assembled, and painted to represent objects other than themselves produces an integral structure in which real and imaginary worlds engage in creative interplay. Cubism provided the first new concept of space since Masaccio – dynamic multiperspective space rather than static uniperspective space.

Action painting strengthened ecological perspective. It abandoned the iconic and symbolic modes of signification that gave meaning to mythological and logical forms of art. Action painting became an art of pure index indicating a direct connection between what was and what is. Its subject matter is the act of painting being self-documented. Jackson Pollock, the foremost action painter, created paintings that presented symptoms of the artist having dripped paint, as well as a documentary map and after-the-fact choreographic score of the movement of his body over a canvas. Pollock was not standing outside an illusionary world looking in on it from a single viewpoint. He firmly rejected a logical single-point perspective. He moved inside, outside, and around a canvas on the floor. He renewed a mythological perspective that ignores a single baseline merging into an ecological perspective in which multiple viewpoints are presented in which top and bottom are eliminated. Pollock wrote: "My painting does not come from the easel ... On the floor I am more at ease. I feel nearer, more part of the painting, since this way I can walk around it, work from the four sides and literally be *in* the painting."[36] Pollock's paintings are dynamic ecosystems of interacting lines, forms, and colors that can be best appreciated from an ecological perspective. Claude Cernuschi writes in his book *Jackson Pollock: Meaning and Significance*:

> What Pollock felt intuitively, that humanity is connected to rather than separate from nature, and the way he depicted this feeling in art, translating the rhythms of nature and man in a single field of energy, also anticipate modern ideas about the origins of the universe.[37]

Pollock's artwork profoundly influenced Allan Kaprow and Robert Smithson, two American artists who extended ecological perspective beyond marks on a canvas out into the audience and into the outdoor environment. Two years after Pollock's tragic death in an automobile accident, Kaprow wrote an article in *ARTnews* titled, "The Legacy of Jackson Pollock,"[38] where he discusses the influence of Pollock's work on his own work as an artist. He proposes that Pollock's works should be viewed in a medium-sized room whose walls are completely covered by paintings. All the paintings together became a total environment surrounding the spectators. The audience finds itself inside the paintings, as Pollock himself had been when he created them. Kaprow extended ecological perspective beyond action painting by inventing Happenings. In action painting the artist is actor; in Happenings the spectator becomes actor. Kaprow wrote in *Artforum*, "There should not be (and usually cannot be) an audience or audiences to watch a Happening."[39] There is a shift from active artist and passive audience to artist and audience as collaborators.

Robert Smithson opened up Pollock's canvas to the outdoor environment with his aerial art and earthworks. Just as Pollock viewed his paintings from above as he made them, Smithson created aerial art that the spectator had to view from an airplane. He defined this kind of art that evolved from his work as a consultant for the Dallas-Fort Worth airport as art that could be seen from an aircraft on takeoff and landing, or not be seen at all. This aerial perspective moves above ordinary eye-level perspective. Smithson, therefore, filmed his *Spiral Jetty* from an airplane, the best vantage point from which to see it. *Spiral Jetty* was a spiral ramp of earth shaped by a bulldozer at the edge of a Utah lake. In addition to seeing it from aerial perspectives, it could also be seen from various ground-level viewpoints. Cernuschi notes:

> The traditional ways of looking at a work of art, either walking before a painting or around a sculpture, suddenly become hopelessly inadequate. The work has no fixed appearance; it constantly changes according to the spectator's relative position inside or outside the work.[40]

The changes in the artwork because of the shifting perspectives of the viewer are augmented by actual changes in the artwork by the forces of nature. It was continually being eroded by wind and water. While filming *Spiral Jetty* from the air with the bright sun glaring up form the Utah lake, Smithson recalled: "My eyes were like combustion chambers churning orbs of blood blazing in the light of the sun. All was enveloped in a flaming chromosphere; I thought of Jackson Pollock's *Eyes in the Heat.*"[41]

Modern art provided Jewish artists with access to an art world that began to honor an ecological perspective. Postmodernism provides an even more comfortable home for Jewish artists to express an integral structure of consciousness. Living in Jewish consciousness precluded entry with integrity into the logical and mythological art worlds with their alien values. Modernism and postmodernism provide opportunities for artistic creation that were absent from Jewish life during the centuries of dispersion throughout logical Europe and the mythological Islamic world. Jews were conceptually alienated and politically excluded from entry into those worlds. In 20th-century America, Jewish artists became active shaping the leading edge of new emerging art forms expressing an ecological perspective compatible with their Jewish consciousness. They became energetically involved in the creation of abstract fields of interacting forms and colors rather than mimetic art, of images, objects, environments, and events perceived from multiple perspectives rather than from a single-point perspective, and of art inviting active participation rather the passive observation. Art curator Ori Soltes observes that "Jewish artists, during the past hundred years or so, have played an inordinately large role – far out of proportion to their numbers in either the general population or the art-making population – in the shaping of modernism in all of its aspects."[42]

Professor of art criticism Donald Kuspit analyzes the art of Mark Rothko and Barnett Newman and draws the conclusion that "their abstract art is quintessentially Jewish – a basic statement of Jewish principle, all the more uncompromising because it operates in a rigorously aesthetic way, without the distraction of Jewish social iconography."[43] He proposes that the expression of the biblical laws of differentiation at the core of Judaism, of what we are expected to do and not to do, is the message in the works of Rothko and Newman. Rothko separates planes from one another and establishes a hierarchy of colors that continue to maintain their differentiation even as they tend to merge. Newman articulates the moment of creation, the separation of light from darkness, by painting a strong line with the energy to separate itself from a uniform background, from an undifferentiated void. Kuspit concludes:

> If I can convince you that the formal structure of the paintings of Rothko and Newman is subliminally Jewish, and in fact conveys in a kind of aesthetic nutshell the basic idea that the Jew represents, then I will have made my point at the deepest level. I will even insist that the art of Rothko and Newman, which is concerned with the aesthetic core of art, uses it to convey the core of Jewishness, that is, it mediates Jewish meaning through the aesthetic immediatism that is inseparable from pure art, suggesting that quintessentializing modernist aesthetic is quintessentially Jewish in spirit, however unwittingly.[44]

In our discussion of globalization, we traced roots of postmodernism in 16th-century Holland with the invention of portable art that could be bought and sold and moved from house to house. Before the Dutch began grinding pigments into oil to paint on canvas stretched over a wooden frame, art was a permanent component of a building. Fresco and mosaic pictures could not be removed from the walls, floors, or ceilings of which they formed an integral part. Vermeer placed the Dutch room inside paintings showing walls with portable paintings hanging on them. This recursive imagery from everyday life where paintings on the wall are in paintings on the wall that can be moved from wall to wall has postmodern overtones.

Modernism removed the picture of the room from the painting leaving the paint itself as the subject. Sometimes, real everyday stuff found its way collaged onto the picture plane. However, the paint and stuff remained trapped on the surface of the rectangular canvas even when the frame was removed. Modern art continued to hang on walls in white box galleries and museums. Postmodernism freed the art from the canvas and moved it off the walls into real space and cyberspace. Art in real space and cyberspace invites people functioning in these spaces to collaborate with the artist in the creative process rather than being passive observers of a finished product.

Gebser's concept of aperspectival art expressing an integral structure of consciousness applies equally to modern and postmodern art. The ecological perspective that emerged in Cubism was expanded in action painting and in conceptual art and is flourishing in postmodern art that is evolving from deconstruction to reconstruction. David Harvey in *The Conditions of Postmodernity* sets modernism and postmodernism in opposition to each other by reprinting Hassan's table "Schematic differences between modernism and postmodernism."[45] Hassan makes two parallel lists of contrary terms: modernism as form (conjunctive, closed) vs postmodernism as antiform (disjunctive, open); purpose vs play; art object/finished work vs process/performance/happening; distance vs participation; centering vs dispersal; transcendence vs immanence, etc. This tabular schema of oppositions attempts to capture ways in which postmodernism might be portrayed as a reaction against the modern. I agree with Harvey's conclusion "that there is much more continuity than difference between the broad history of modernism and the movement called postmodernism [...] The 'structure of feeling' in both the modern and postmodern periods lies in the manner in which these stylistic oppositions are synthesized."[46]

From deconstruction to reconstruction

The postmodern reaction against modernism is the early deconstructive stage, a transitional stage towards achieving integral consciousness. The more mature stage of postmodernism is a reconstructive stage where a creative synthesis between the opposites is achieved through dynamic interplay between them. It is as if postmodern deconstruction disintegrates earlier art forms freeing up their elements to be reconfigured in new ways in an ongoing process.

Jacques Derrida proposes deconstructionism as a postmodern way of reading texts that goes beyond decoding a message to ceaseless questioning of interweaving texts through thoughtful play with contradictory messages and multiple references. This kind of play in earnest creates opportunities for popular participation and democratic determinations of alternative viewpoints.

> The deconstructivist impulse is to look inside one text for another, dissolve one text into another, to build one text into another. Derrida considers, therefore, collage/montage as the primary for m of postmodern discourse [...] The cultural producer merely creates raw materials (fragments and elements), leaving it open to recombine these elements in any way they wish. The effect is to break (deconstruct) the power of the author to impose meanings or offer a continuous narrative.[47]

Artist Lowry Burgess extends Derrida's concept of deconstruction in reading literature into reconstruction in the visual arts. Burgess, the creator of the first artwork launched by NASA to circle our planet and art professor at Carnegie Mellon University, contends that the earlier art forms like painting and sculpture do not live comfortably in the postmodern reality.

> They want to be more dynamic, more rapid, more explorative, and inquisitive; they want to be more democratic, more synaesthetic, more polyvalent, more free-associative. In other words, this new framework demands a de-gestalting and fragmentation – then a re-assembly on another plane of consciousness.[48]

This de-gestalting and re-assembly at another plane is analogous to what Nobel laureate Ilya Prigogine calls "dissipative structures" in nature. Dissipation, fragmentation, deconstruction of a natural system into a chaotic state frees up its elements and make them available for reordering, restructuring, reconstructing themselves into a new structure. He defines dissipative structures in his book *Order out of Chaos: Man's New Dialogue with Nature*:

> We now know that far from equilibrium, new types of structures may originate spontaneously. In far-from-equilibrium conditions we may have transformation from disorder, from thermal chaos, into order. New dynamic states of matter may originate, states that reflect the interaction of a given system with its surroundings. We have called these new structures *dissipative structures* to emphasize the constructive role of dissipative processes in their formation.[49]

The biblical narrative offers us an image of the ultimate deconstruction of a text and its reconstruction at a different level of consciousness. Moses took the "Made by God" stone tablets with the text of the Ten Commandments and smashed them to bits. The "Made by God" text was deconstructed so that it could be reconstructed by a human hand. The Torah is received to this day written by the hand of a scribe on a spiral scroll rather than engraved by God on stone tablets. In the Talmud, we are told that Moses was praised for his physical act of deconstruction to free the letters from closed rectangular forms to be reformulated by human hands on spiraling scrolls linking the end to the beginning in an endless flow.

It is told that the Hebrew letters from the broken tablets were scattered over the desert to invite us to gather them and re-assemble them to recreate the text as a message with fresh meanings for every generation. Artist Miriam Benjamin and I made all the letters appearing on the tablets in the biblical text from clay dug from the Negev desert that flows into the Sinai. We scattered the letters throughout the Sinai desert from the Red Sea coast to the mountainous regions. We attached a card to each letter inviting the finder to incorporate it in a reconstructed biblical message.

The traditional Jewish method of reading biblical texts is a far more extensive deconstructive process than Derrida discusses. It moves beyond deconstruction as alternative readings of texts to morphological dissection that frees up the letters forming words to be reconfigured in ways that create fresh meanings. It even extends to subatomic deconstruction of texts by finding significance in the shapes of the Hebrew letters and configurations of their parts. The original Hebrew Torah text written as lines of consonants without vowels and punctuation calls attention to each letter and invites multiple readings of words and sentences by its ambiguity. It is as if the Bible begins, "n th bgnnng gd crtd th hvns nd th rth." If a single one of the 304,805 letters in the hand-written Torah scroll is missing, incomplete, or if ink chips off one small part of a letter, then the entire scroll is unfit for public reading.

There are a number of traditional systems for seeing through the text rendering it diaphanous through play with letter relationships and combinations. Rabbi Moshe Cordevaro, a central figure in the group of kabbalists living in 16th-century Tzfat, Israel, wrote: "Knowledge of our holy Torah's secrets can be discovered through combinations: numerical letter values (*gematria*), letter permutations, exchanging first with last letters (*ATBaSh*), shapes of letters, first and last verses, skipping letters (ELS codes), and letter combinations (*notaricon* acronyms)."

The Talmud teaches that there are seventy different viewpoints from which to understand every biblical verse. The playful spirit in which sacred scripture is approached is described in the introductory chapter. Four levels of reading a text in the acronym *PaRDeS* are the simple literal meaning of the words (*P'saht*), hints of innate significance (*Remez*), homiletic interpretations (*Drosh*), and the mystical meanings hidden in the white spaces between the black letters (*Sod*). Through *PaRDeS*, we learn that Jacob's ladder was a spiral staircase for bringing Torah from heaven to earth as a metaphor for Mount Sinai from the Hebrew words for "ladder," "spiral," and "Sinai" all sharing the numerical value of 130. Since each Hebrew letter has a numerical equivalent, we can understand a text by calculating the values of words and relating them to each other. This numerical method of reading is called *gematria*.

The *ATBaSh* system replaces the first letter of the Hebrew alphabet *alef* with the last letter *tav*, the second letter *bet* with the next last letter *shin*, and so on. An English equivalent would be called *AZBY* with *z* replacing *a*, *y* replacing *b*, *x* replacing *c*, and so on. It is used in the Bible itself when the word for Babylonia, *BBL* (*bavel*) is also found in the same passage in its *ATBaSh* form *ShShKh* (*sheshakh*). "How has Sheshakh been captured and the glory of the whole world seized? How has Babylonia become a desolation among the nations?" (*Jeremiah 51:41*). The use of *ATBaSh* by the Bible itself gives license to using it freely in reading any other portions of the biblical text. A divine commandment is a *mitzvah* spelled *mem*, *tzadi*, *vav*, *hei* (*MTzVH*). The *ATBaSh* for *mem* is *yod* (*Y*) and for *tzadi* is *hei* (*H*), thereby creating the unpronounceable four-letter name of God (*YHVH*). God and His commandments are unified by *ATBaSh*.

The Israelites were commanded to write the words of the Torah on parchments affixed to the doorposts of their homes (*Deuteronomy 6:9*). Until today, cases holding these mini-Torah parchment scrolls could be seen on the doorposts of Jewish homes. They are called *mezuzot*, derived from the root word *zaz*, meaning movement. On each scroll, a scribe hand writes the biblical passages from *Deuteronomy 6:1–9* and *11:13–21* beginning "Hear, O Israel, *YHVH* is our Lord, *YHVH* is One." The scroll is rolled up with the biblical passages on the inside of the scroll. On the reverse side of the scroll, the divine name *YHVH* (meaning "Is-Was-Will Be") is written as *KUZU* to demonstrate the *YHVH* cannot be experienced as a static object, but rather as a dynamic process. *KUZU* moves the divine name one letter forward. It is spelled with each of the letters that follow *YHVH* in the Hebrew alphabet. It is as if we were to write GOD as HPE, H being the letter following G, P the letter following O, and E the letter following D. In addition to moving each of the letters in *YHVH* forward, *KUZU* is written upside-down to invite us to learn the Torah from multiple vantage points.

Reversing the order of letters can reveal relationships between material and spiritual realms. The Hebrew word for "material" is *homer* spelled *HMR*. The word for "spiritual" is *ruah* spelled *RUH*. The four letters formed by the lips when spoken, *BUMF*, are interchangeable. When we eliminate the interchangeable letters *M* and *U*, then we can read *H-R* and *R-H* as the same word backwards to learn that the difference between the material world and the spiritual world is one of perspective. Both worlds are unified by a shift in our angle of vision. We experience the world as being spiritual by the quality of our perception of the material world.

The shape of each Hebrew letter carries profound messages. Unlike English letters written sitting on a line below them, a Hebrew scribe suspends letters from a line above. This top line from which the Hebrew letters hang symbolizes the boundary between the writing and ideas that exist beyond the text. None of the Hebrew letters breaks through the barrier except the letter *lamed*. The name of the letter *lamed* also means "to learn" and is the root of Hebrew words related to studying and teaching as in *Talmud*, the name of the major compilation of Jewish creative thought studied by Jews worldwide. Learning in Judaism invites breaking through the barrier and soaring to new heights of insight and understanding. Unlike the English term "educated man," which in the past tense tells us that learning has been completed, the comparable Hebrew term *talmid hakham* means "a wise learner," one who is forever a student whose learning never ceases to break through barriers to higher levels of knowledge.

The biblical creation narrative in *Genesis* contains all the letters with the exception of the *samekh*. The shape of the *samekh* resembles a zero. Since it is the only one of the 22 Hebrew letters that has a closed form, it cannot describe the dynamic, open-ended, ongoing creation of an expansive and expanding universe. Indeed, *samekh*, excluded from the creation narrative, is the initial letter of the Hebrew word for "closed" (*sagur*).

Reading biblical words as acronyms is a method called *notaricon*. For example, the Bible's central affirmation of monotheism, "Hear, O Israel, God is our Lord, God is One" (*Deuteronomy 6:4*), can be read as synaesthetic integration of auditory and visual senses by linking this verse to a second biblical verse, "Raise your eyes on high and see Who created these" (*Isaiah 40:26*). These two biblical portions are read together on the same Sabbath in synagogues. The Hebrew word for "Hear" is ***SheMA***, spelled *Shin*, *Mem*, *Ayin*. It is an acronym for ***S****e'u* ***M****arom* ***A****ynaykhem*, "Raise your eyes on high." We learn from linking these two verses that we are to experience the one and only God through multiple senses.

"ELS (equidistant letter sequence) codes" is a method of reading the Bible by skipping an equal number of spaces between letters in the text to discover a new word. Rabbi Weismandl found that if we start with ***Tav***, the final letter of the first word of the Bible, *b'reshi**T*** (in the beginning), and count every 50th letter in the book of *Genesis*, we read the word *Torah*. If we start with the first *tav* in the next book of the Bible, *Exodus*, and count every 50th letter we will once again read the word *Torah*. In the last two books of the Pentateuch, *Numbers* and *Deuteronomy*, the word *Torah* is found backwards, facing towards the center with equidistant letter sequences. Inspired by Rabbi Weismandl's discovery before the digital age, mathematics professor Eliyahu Rips at the Hebrew University of Jerusalem and Doron Wiztum and Yoav Rosenberg of the Jerusalem College of Technology studied this phenomenon scientifically with the aid of computers. They discovered in the Torah numerous hidden messages encoded in equal letter sequences. After passing the peer review of three skeptical reviewers, the leading American mathematics journal *Statistical Science* published their findings, "Equidistant Letter Sequences in the Book of Genesis."[50] The most stringent standard applied in scientific studies is 1 in 1,000. Rips discovery of hidden messages in the biblical text encoded in equal letter sequences using digital quantitative tools resulted in levels of significance of 1 in 200,000.

I heard Rabbi Simcha Hochbaum of Hebron tell the story of the ultimate deconstruction of the biblical text. He told of two great Torah scholars who grew up as inseparable friends. They were learning partners from childhood through their rabbinic studies, parting only after being appointed to posts as rabbis in two separate towns. Every Friday each sent the other a letter that became the basis for their Sabbath sermons the next morning. A faithful messenger traveled two hours each way on horseback braving highwaymen, stormy weather, and muddy dirt roads for a number of years. One day, a fierce gust of wind sent his rabbi's letter flying out of the pouch on the side of his horse. He immediately dismounted to chase the letter. The wax seal on the envelope broke and the letter flew out. As he retrieved the letter, he glanced at it and was shocked to see that it was a blank sheet of paper. "Why would his rabbi send an empty page to his colleague?" he asked himself. He stuffed it back into the envelope and delivered it. As he left with the return letter, curiosity overwhelmed him and he opened it only to find it too was a blank sheet of paper. Infuriated that he had braved all sorts of dangers to faithfully deliver empty pages, he confronted his rabbi as he handed him the blank sheet. The great rabbi explained that the deepest secrets of the Torah are found in the white spaces between the Hebrew letters. He assured his faithful messenger that he and his beloved friend could only communicate the most profound ideas with each other by writing with the white spaces.

The definitive deconstruction of a text is to reveal the spaces between the letters and be able to reconstruct the text and decipher its deepest meaning reading a blank sheet of paper. When we look at the negative space around the first letter of the Bible, *bet* (in), we see the letter, *pei*, which means "mouth." Making a figure-ground shift in perception, we see the *bet* as the inside of the *pei*. When a scribe writes a *pei* in black ink, the white space within the *pei* is the shape of a *bet*. We read the words "In the beginning God created…" starting with a black *bet* that emanates from within a white *pei*, the mouth of God. We begin reading the Torah from the white space preceding the black ink.

This Hasidic tale has its parallel in contemporary arts. Blank white canvases presented as artworks inspired the presentation of silence as a music composition in which a musician observed a blank score and played nothing. Robert Rauschenberg exhibited sets of rectangular canvases that he had painted entirely white. His works are titled: *White Painting [one panel], White Painting [three panel], White Painting [four panel], White Painting [seven panel]*. Rauschenberg deconstructed the text of Western art to the zero point eliminating all imagery thereby freeing all imagery for recombination in infinite numbers of new relationships. He became the master of collage, juxtaposing images from radically different sources on his white canvases. He created dynamic tensions between the appropriated images with expressionist strokes of raw colored paint reconnecting pictures from popular culture to the tradition of painting.[51] Inspired by Rauschenberg's white paintings, composer John Cage created his silent piece *4'33"*, a work in three movements. A year after Rauschenberg's show, his *4'33"* was "played" by a pianist in a concert hall seated quietly at a piano for four minutes and 33 seconds signaling the beginning and end of each movement by opening and closing the keyboard cover.[52]

Morphological analysis of visual culture: biblical fringes

The visual commandment (*mitzvah*) dealing with seeing fringes flowing out of the four corners of a garment to remind us not to stray after our eyes is the exemplary visual *mitzvah* from among the 613 commandments found in the Bible. I will analyze morphologically how these biblical fringes in the visual culture of Jewish praxis express the structure of Jewish consciousness and inspire expression in contemporary art. This visual *mitzvah* in Jewish praxis is derived from the biblical passage *Numbers 15:37–41*:

> God said to Moses, saying: "Speak to the Children of Israel and say to them that they shall make themselves fringes (*tzitzit*) on the corners of their garments, throughout their generations. And they shall place upon the fringes of each corner a blue thread. It shall constitute fringes for you, that you may see it and remember all God's commandments and perform them and do not stray after your heart and after your eyes that you may remember and perform all My commandments and be holy to your God. I am the Lord your God who brought you out of the land of Egypt to be a God to you. I am the Lord your God."

The biblical Hebrew word for fringes is *tzitzit* from *tzutz* meaning "to look." *Tzitzit* are something to look at. We look at fringes tied with 5 knots, 39 spirals, and 8 branches, breaking out of the four corners of a rectangular garment. Since contemporary clothes are not rectangles, Jews don rectangular prayer shawls (*talit*) and undergarments (*talit katan*) in order to perform the *mitzvah* of *tziztit*. The importance of the *mitzvah* of *tzitzit* is emphasized by the placement of the passage from Numbers on *tzitzit* in a major position in the liturgy attached to two passages from Deuteronomy that begin with the central attestation of Jewish faith: "Hear O Israel, *YHVH* is our Lord, *YHVH* is One."

The significance of the visual *mitzvah* of *tzitzit* in the visual culture of the Jewish people relates to to form, number, and process. These three parameters of biblical visual culture as enunciated in ancient kabbalistic literature parallel both our contemporary scientific understanding of the universe and future directions for the visual arts. *Sefer Yetzirah* (*The Book of Creation*)[53] often attributed to the patriarch Abraham, begins: "The universe was created with three *SiPuRim*, with *SePheR*, with *S'PhaR*, and with *SiPuR.*" In Hebrew, words for form *SePheR*, number *S'PhaR*, and process *SiPuR* share the same *SPR* root that is revealed in the word for SPiRal in many languages, ancient and modern, and in the English words SpiRitual and inSPiRation.

The kabbalistic morphology of form, number, and process in spiritual structures finds its scientific parallel in the structures of natural systems. D'Arcy Thompson's classic work, *On Growth and Form*,[54] demonstrates how living organisms are shaped by form, number, and process. In *The Sand Dollar and the Slide Rule: Drawing Blueprints from Nature*,[55] Delta Willis extends Thompson's research to explore the relationships between natural systems and the human design process. Judaism considers the Bible to be the blueprint for the creation of the material world. Therefore, it is reasonable to see structures in nature parallel structures revealed through Torah study and Jewish praxis. The spiral form of the galaxies and DNA can be seen in the double windings of the Torah scroll. The qualitative differences between the 92 elements are a manifestation of atomic number paralleling the numerical values of Hebrew words that reveal hidden meanings. Photosynthesis, the most significant chemical process in the maintenance of life on our planet, is analogous to the biblical narrative that relates life enhancing processes that lead from slavery to freedom and from narrowness of thought to ecological perspective.

Morphological analysis of *tzitzit* in terms of form, number, and process reveals their significance in the development of Jewish consciousness. The Israelites where enslaved in the *malben*, both brickyard and rectangle. The biblical passage on *tzitzit* ends with liberation from Egypt's narrow straits. *Tzitzit* break open the corners of a rectangular garment to let energies flow from the narrow confines of a closed frame. A garment not having four corners does not require *tzitzit*. In tying *tzitzit*, four strings are laced through a hole made at each of the corners so that we see eight strands. A double knot is tied integrating all eight strands. The double knot is analogous to the two intertwined triangles, one pointing upwards and the other downwards, that constitute the Star of David. It teaches that spiritual ascent must be integrated with the lowering of spirituality down into everyday life. This is the first of five double knots that separate four sets of spirals from each other and the long flowing branches at the end of the fringe. Spirals and branches are open-ended forms like the alternative patterns of growth in palm and cedar trees. The psalmist uses this tree metaphor to elucidate the moral goal of Judaism, "A righteous person will flourish like a date palm, like a cedar of Lebanon he will grow tall" (*Psalm 92*). It is the highlight of the week to sing this psalm, "A song for the Sabbath day," in the synagogue. The spirals and branches of the *tzitzit* remind us to liberate ourselves from narrow patterns of thought to reach righteousness through the open-ended flow seen in the growth patterns of trees. The eight free-flowing branches hang down like roots to the upward branches of the *menorah*, the most ancient symbol of Judaism. They must be twice as long as the four sets of spirals contained between five knots. The integrating structure of the knots and the growth structure of the spiral turns give way to free flowing branches, the longest portion of the fringe.

According to the biblical passage on *tzitzit*, one of the strands of each fringe should be sky blue. The sages of the Talmud tell us, "The blue wool resembles the ocean, the ocean resembles the color of the sky, the sky resembles the purity of the SaPphiRe, and the sapphire resembles the divine throne."[56] The special blue dye (*t'khelet*) needed for coloring the *tzitzit* was derived from a Mediterranean sea snail. In the period of the Babylonian exile after the destruction of the Temple in Jerusalem when Jews lived far from the sea, the blue thread went into disuse since this required dye could not be obtained without great effort and expense. After more than two millennia, dying one strand blue is being revived in Israel today where the rediscovered spiral sea snails of ancient times are grown in salt-water tanks. The blue and white color of the Israeli flag is based upon the blue and white strands of the *tzitzit*. The blue thread linking sky to ocean is a metaphor for the religious obligation of bringing heaven down to earth. When an artist mixes sky blue pigment with earth red pigment (the Hebrew word for red is *adom* sharing the same root as *adamah*, earth, and *adam*, humanity) the resultant color is purple. The Bible (*Exodus 19:5* and *Deuteronomy 6:7*) calls the Jewish people "*am segulah*" (*SeGuLah* is related to the word for purple *SeGoL*), a nation chosen to teach ways of drawing spiritual realms above into every aspect of down-to-earth living.

Looking at *tzitzit* flowing from the four corners of a *talit* reminds us not to encounter the physical world without transforming it into a spiritual one. The 613 biblical *mizvot* are divided into 248 positive and 365 negative *mizvot*. The positive *mizvah* of *tzitzit*, like tying a string around a finger to remember what to do, reminds us to observe the biblical commandments. The negative *mitzvah* not to stray after what our eyes and hearts desire relates to not acting out our animal instincts divorced from deeper values, not admiring beauty of form separated from appreciation for goodness, and not dissociating material and spiritual realms from each other.

The Talmud[57] relates a story to teach how observing this visual *mitzvah* can keep us from lusting after surface gorgeousness separated from inner beauty. A yeshiva student careful to observe the *mitzvah* of *tzitzit* traveled to a port city where he heard there lived an amazingly gorgeous harlot. He paid her attendant 400 gold pieces and entered a room sensuously set with six beds of silver and the highest one of gold. She disrobed and sat herself nude on the golden bed. When the man began to undress, his four *tzitzit* flew up and slapped him in his face. As he began to run out, the woman cried out, "By Jupiter! I will not let you leave me until you tell me what you find wrong with me." The man answered, "I have never seen a woman as beautiful as you but the *tzitzit* that my God has commanded me to wear appeared to me like four witnesses to testify against my sin." The woman sold all her possessions except for her gold bed and traveled in search of the man who had had the strength to resist taking her beautiful body. When she found his yeshiva, she told the story to the rabbi who headed the yeshiva and begged him to teach her. She learned the Torah with love and enthusiasm and converted to Judaism. The rabbi then said to her, "Go, my daughter, and take what is yours. The same bed that you used for sin, you may now enjoy in a permissible manner." The yeshiva student happily married a most stunning bride whose outer charms were now matched by the beauty of her soul.

Tzitzit also communicate spiritual messages through numbers. The four sets of spirals held together by 5 knots wind down in a sequence of 7, 8, 11, and 13 turns (according to the Ashkenazi tradition). Seven days of divine creation is followed by the eighth day in which humanity joins with God in continuing the creation. 7 + 8 = 15, the numerical equivalent of *YH*, the first two letters in the divine name. The numerical value of second two letters, *VH*, is 11. The full divine name *YHVH* equals 26. The fourth set of 13 turns is the numerical value of *ehad*, the Hebrew word for "one." Looking at the *tzitzit*, we read, "God is One." God (26) plus One (13) equals 39, the numerical equivalent of *KUZU*, the divine name *YHVH* in motion. *YHVH* is the divine name associated with *tiferet*, the inner beauty in the flow of divine light down into the material world. Writing *YHVH* with the next Hebrew letter in the alphabet represents the dynamic quality of this flow. As we mentioned above, it is as if we were to write the word GOD as HPE (H follows G, P follows O, and E follows D). Like *KUZU*, HPE would set GOD in motion. Furthermore, the numerical value of the word *tzitzit* is 600 (*tz* = 90, *i* = 10, *tz* = 90, *i* = 10, *t* = 400). Coupled with the 8 strings and 5 knots, we arrive at 613, the number of Torah injunctions.

Before donning the *talit* every morning, I recite from *Psalm 104*: "Bless *YHVH*, O my soul; *YHVH*, my God, You are very great; You have donned majesty and splendor, cloaked in light as with a garment, stretching out the heavens like a curtain." I then express my readiness to wrap my body and soul in "the illumination of the *tzitzit* that has the numerical value of 613." I continue:

> Through the commandment of *tzitzit* may my life force, spirit, soul, and prayer be rescued from external forces. May the *talit* spread its wings over them and rescue them like an eagle rousing his nest fluttering over his eaglets. May the commandment of *tzitzit* be worthy before the Holy One, Blessed be He, as if I had fulfilled it in all its details, implications, and intentions, as well as the 613 commandments that depend upon it. Amen, Selah!

I visually examine the *tzitzit* as I move my fingers through them and then wrap the *talit* over my head covering my whole body with it while reciting from *Psalm 36*:

> How precious is Your kindness, O God! As human beings take refuge in the shadow of Your wings, may they be sated from the abundance of Your house and may You give them drink from the stream of Your delights. For with You is the source of life – by Your light we shall see light.

At a high point in the morning prayers, I gather together all four *tzitzit* in my left hand while saying: "Bring us in peacefulness from the four corners of the earth and lead us with upright pride to our land." I cover my eyes with my right hand and recite, "Hear O Israel, *YHVH* is our God, *YHVH* is One." I continue reciting two paragraphs from *Deuteronomy* that follow from this central affirmation of Judaism. When I come to the next paragraph from *Numbers* about *tzitzit*,I hold the four fringes together in both hands and kiss them each of the three times the word *tzitzit* is mentioned. I look closely at the gathered fringes when saying "that you may see it and remember all God's commandments and perform them and do not stray after your heart and after your eyes." At this point, Jews from Islamic lands will often take the fringes and touch them to both closed eyes transforming the visual sense to a more intimate tactile one. After reciting the biblical paragraph on *tzitzit*, the gathered fringes are released while saying: "His words are living and enduring, faithful and delightful forever and to all eternity."

My description of the daily practice of wearing, holding, kissing, and looking at *tzitzit* is one aspect of the transition from form and number to process. Another aspect of that transition to process can be read from the textual context of the biblical passage on *tzitzit* in relation to the biblical narrative. The biblical passage on *tzitzit* (*Numbers 15:37–41*) is the last paragraph of the Torah portion that begins with the narrative about the Israelites spies sent from the desert to gather intelligence about the land of Canaan (*Numbers 13:1–33, 14:1–45*). "The men headed north and explored the land, from the Tzin Desert all the way to ReHoV" (*Numbers 13:20*). Note that *RHV* is the root of the word for "wide-open expanses." Leaving Egypt, the height of built civilization, to the Sinai desert automatically liberated the Israelites from the narrow straits of an enslaved people to the freedom of the broad expanse of the desert landscape. When I drive from my home in a suburb of Tel Aviv to visit my son and his family in the Negev desert, I can sense this biblical transition. Looking out of my apartment windows, I only see other apartment buildings. Leaving the narrow canyons of a vertical built environment and heading into the desert mountains, I encounter unrestrained horizontal vistas.

Ten of the twelve spies returned with disparaging reports about the land of Canaan. They opposed entering the land preferring to remain in the desert. In the desert, they could enjoy a spiritual life of learning the Torah without the work and worries of earning a living. All the Israelites' needs were met. Their food was the manna that fell from heaven and water came from Miriam's well. They feared entering the land of Canaan for it would rob them of their spiritual existence and make them toil as they had as slaves in Egypt. Why was it such an unforgivable sin to prefer a purely spiritual life to one of mundane tasks and hard work? The spiritual leaders of ten of the tribes were condemned to death in the desert for their grave transgression.

"Among the men who had explored the land, Joshua son of Nun and Calev son of Yefuneh tore their clothes in grief. They said to the whole Israelite community, 'The land through which we passed in our explorations is a very, very good land … a land flowing with milk and honey'" (*Numbers 14:6–8*). Joshua, the faithful disciple of Moses, had no interest in rebelling against God's will as communicated through Moses. His good report earned him the privilege of entering the Promised Land. God said, "The exception is my servant Calev who showed *a different spirit* and followed Me wholeheartedly. I will bring him to the land that he explored, and his descendents will possess it" (*Numbers 14:24*). In our day, the descendents of Calev of the tribe of Judah are being ingathered from the four corners of the earth to Israel, the land that Calev was able to see as being very, very good. Most of the Jews in Israel today descend from the tribe of Judah. Most descendents of those who feared entering the Promised Land after the exodus from Egypt are the lost tribes.

What was Calev's different spirit? What could he see that the others could not? Rabbi Schneerson explains that Calev understood that the other spies were wrong in believing that spirituality flourishes best in seclusion and withdrawal, in the protected isolation of the desert.

> The purpose of life lived in Torah is not the elevation of the soul; it is the sanctification of the world. The end to which every *mitzvah* aims is to make a dwelling place for G-d in the world – to bring G-d to the light within the world, not above it. A *mitzvah* seeks to find G-d in the natural not the supernatural. The miracles which sustained the Jews in the wilderness were not the apex of spiritual existence. They were only a preparation for the real task: taking possession of the Land of Israel and making it a holy land.[58]

Rabbi Kook[59] shares the view that the spies were severely punished because they could not see how to fuse the physical with the material in the Land of Israel. In the Land of Israel, we have the possibility of proving that mundane tasks can be sanctified and part of the divine service. Calev could make a perceptual shift and see that building the land and tilling the soil could be transformed into acts of spiritual significance.

During the Nazi conquest of Europe, Nazi troops attempted to humiliate a great rabbinic leader in front of his congregation by forcing him at gunpoint to sweep the streets, collect garbage, and march waving the Nazi flag. As one of the few survivors of the Holocaust in his family, he was fortunate enough to begin a new life in Tel Aviv. Each year, the week before Israel's Independence Day, he would rise early to join sanitation workers in sweeping the streets and collecting garbage. On Independence Day, he marched in Tel Aviv waving the Israeli flag. Being free to keep his city clean and wave his flag transformed humiliation into an expression of pride. The sin of the spies was their inability to see that the very same activities that they were forced to endure as slaves in Egypt could be transformed into holy work when performed freely of one's own volition.

Calev had "another spirit" able to conceptualize bringing the expansive experience of the desert into settlements in the Land of Israel. The spies who left Egypt (*MiTzRayim* = narrow straits) and crossed the desert entered the Promised Land at the Tzin Desert with the charge to reach Re<u>H</u>oV (*Numbers 13:21*). They failed to arrive at Re<u>H</u>oV, which alludes to wide expanses. Forty years later, Joshua sent spies who found refuge in Jericho with the woman innkeeper Ra<u>H</u>aV, whose name means expansiveness (*Joshua 2:1*). On holidays and each celebration of the new month, Jews sing King David's words: "From the narrow straits (*haMaTzaR*) did I call upon God; God answered me with expansiveness (*meR<u>H</u>aV*)" (*Psalm 118*).

The Torah portion telling the story of the spies ends with the passage about *tzitzit*. Looking at the *tzitzit*, we see flowing tassels break open the closed rectangle at its four corners. We see four knots open up to spirals symbolizing open-ended growth and a fifth knot open up to free all eight strands to flow outwardly in expansive branching. These branching strands streaming downward from the corners of a *talit* symbolize the flow of divine light down to earth.

Art as visual *midrash*: *Four Wings of America*

Midrash is 2,000 years of creative narratives designed to elucidate biblical concepts. It takes the biblical narratives themselves and spins out tales that read between the lines of the biblical text and that reveal messages hidden in the white spaces between the Hebrew letters. These inspirational stories form a vast literature illuminating biblical texts from countless alternative viewpoints. Postmodern art provides a context in which traditional story telling can be transformed from a verbal activity into visual one. "In a postmodern framework, art is once again about something beyond itself; it defines a particular narrative or world view."[60] Exemplary narrative art can be seen in the artworks of the husband-wife teams Christo and Jean-Claude and Newton Harrison and Helen Mayer Harrison. Their artworks are narratives of their dialogue with social, political, and environmental factors presented through the interplay between of visual and verbal modes of expression.

New narratives emerging from a visual artist's dialogue with biblical texts can be called "visual *midrash*." My visual *midrash*, *Four Wings of America*, creates an innovative form of Jewish visual culture that acts in counterpoint to the age-old ritual drama of donning a *talit* with *tzitzit* flowing from its corners. I extend the postmodern pattern of art as the narrative of an artist's dialogue with social, political, and environmental realms into spiritual and ethical realms as my New World tale, *FourWings of America*, extends into an Old World tale, *Fringed Hut in Munich*. The story of *tzitzit* flowing from the four corners of America continues in a visual *midrash* in which *mega-tzitzit* emerge from four corners of a giant *talit* in Munich constructed from planks of pinewood supplied by BMW for "Sky Art '83." The giant *talit* is built as a *sukkah*, the traditional hut that is both a reminder of the desert dwellings of the Israelites during their exodus from Egypt and an invitation to initiate an era of world peace.

Often *midrash* forges creative connections between a word in the Bible that appears several times in different contexts. In *Numbers* we read, "Speak to the Children of Israel and say to them that they shall make themselves fringes on the *corners of* (*kanfai*) their garments, throughout their generations … I am the Lord your God who brought you out of the land of Egypt." Before the Israelites received the Ten Commandments, God tells Moses, "Tell the Israelites: 'You saw what I did in Egypt, carrying you on *wings of* (*kanfai*) eagles and bringing you to me" (*Exodus 19:4*). Forty years later standing on the east bank of the Jordan River, Moses reviews the laws of the Torah for the generation born in the desert before they enter the Promised Land. He said, "Make yourself fringes on the four *corners (kanfot)* of the garment with which you cover yourself " (*Deuteronomy 22:12).* Before donning his prayer shawl each morning, a Jew says, "May the *talit* spread *its wings* (*kanfav*) [...] like an eagle rousing his nest, fluttering over its eaglets." The biblical prophesy, "He will ingather the dispersed ones of Judah from the four *corners* (*kanfot*) of the earth" (*Isaiah 11:12*), is being realized in our day. The biblical Hebrew word used for the four "corners" of one's garment and metaphorically as the four "corners" of the earth is the same word that is used for "wings." The foremost biblical commentator Rashi (11th-century France) points out the links between corners and wings, "The *tzitzit* are placed 'on the corners (*kanfot*) of their garments,' alluding to God having freed the Israelites from Egypt, as it states, 'and I carried you on the wings (*kanfot*) of eagles.'"[61]

Four Wings of America is a visual *midrash* that conceptually links corners of a garment to corners of the land to wings. It was one of twenty artworks that my wife, the artist Miriam Benjamin, and I created as part of the official celebration of Miami's centennial. When we moved to Miami from New York, we sensed that we had moved to one of the four corners of America. These artworks explored relationships between the four corners of continental United States and its geographic center. We made large white rope *tzitzit* with a sky blue thread with the thought of attaching them to the four corners of America. Since corners are wings in biblical Hebrew, we invited American Airlines, the largest US corporation in the wing business, to sponsor our artwork. We placed large rope *tzitzit* on the boardroom table to explain to the airline executives their ritual significance and why we wanted to create a visual *midrash* by placing them at the four corners/wings of America. It became apparent that our proposal was appreciated, when one of them said, "It is as if the United States is spiritually lifted up by its four corners as the blue thread of the fringes links the sea to the sky." They agreed to sponsor the project and flew Miriam and me to the four corners of America to physically realize our spiritual metaphor. Since American Airlines is the only airline with non-stop flights from Miami to Seattle, its public relations people were pleased with the concept.

We drove from Seattle to Neah Bay, an Indian reservation at the end of the Olympia Peninsula in Washington State, attached the *tzitzit* to a tree at the shoreline. The *tzitzit* flowing outward into the Pacific Ocean transformed the northwest corner of the continental United States by their presence. At the southwest corner, the *tzitzit* shuddered in the wind hanging from to the steel wall that separates San Diego from Tijuana at the Pacific Ocean. *Tzitzit* flowed into the Atlantic Ocean from huge barnacle-encrusted boulders on the Maine coast and from swaying palms shading the beach of a balmy Florida bay.

Biblical passages on *tzitzit* linking them to the exodus from Egyptian bondage invite us to appreciate our freedom. The sky blue strands of *tzitzit* flowing freely from the four corners of America also tell America's story that links the heavenly blessing of freedom to the oceans crossed by those yearning to be free in the New World. At the request of the Continental

Congress, Thomas Jefferson, Benjamin Franklin, and John Adams proposed a seal for the newly independent United States of America that shows the Israelites escaping to freedom from Egyptian bondage through the divided waters of the Red Sea while Moses stood on the shore with his hand held high over the sea. President George Washington repeated the same biblical message of freedom in his letter to the Jewish community of Savannah. His letter quoted in the introductory chapter, draws the parallel between God's delivering the Hebrews from oppression in Egypt to the freedom of the Promised Land and the providential agency in establishing the United States separated from European oppression by a vast sea. He prays that the same wonder-working Deity that freed the Israelites from slavery in Egypt would "still continue to water them with the dews of heaven" as a Jewish community living in freedom in America.

In synagogues in each of the four corner cities – Miami, San Diego, Seattle and Portland (Maine) – I participated in the weekday morning services wearing *tzitzit* flowing out of the four corners of my *talit*, a white woolen rectangular shawl with a series of stripes on both ends like giant bar codes. The stripes are parallel to call attention to the multiple paths of the twelve Israelite tribes, each representing different personality traits and alternative viewpoints. I photographed the spontaneous groupings of men in striped shawls as they gathered around the Torah scroll to kiss it as it was carried from the ark to the reading table. It brought to mind the herds of zebras in a National Geographic film I had seen. The zebras gathered together for protection. However, when a zebra was about to give birth she separated herself from the herd so that her unique stripe pattern would be imprinted on the newborn's mind. If the newborn zebra were to first see the patterns on other zebras, it would be unable to identify its mother in the herd for nursing and would die of starvation. Like a bar code that identifies a product, zebra stripes serve a biological survival function of imprinting the identity of a particular zebra as mother. Perhaps those Jews who come together each morning donning a striped *talit* and seeing the *tzitzit* will never forget their identity. I photographed zebras in the zoos of each of the four corner cities and juxtaposed them with the photographs of the men in striped shawls.

I painted a mural incorporating the bar code stripes from the cover of *ARTnews* magazine on a wall in downtown Miami with the caption, "We stand illiterate before bar codes that supermarket lasers read with ease." Seeing the common pattern in the secret language of our digital age and in zebras' survival mechanisms and the ritual act of wearing a *talit*, attests to the grand ecosystem in which all is interconnected in divine oneness. The integral structure and ecological perspective that characterizes Jewish consciousness are playfully revealed through narrative artworks.

After attaching *tzitzit* to the four corners of America, I sensed that I needed to experience the center. Lebanon, Kansas, is the geographical center of continental United States. I flew to Kansas City and took a small plane to Selena, rented a car and drove through miles of corn fields to Lebanon, a town of 350 souls in the center of the northern tier of Kansas near the Nebraska border. Shortly before arriving at Lebanon, I passed through a town with a sign on its main street, "The Largest Ball of Twine in the World." I pulled over to see a ten-foot high ball of string. It looked to me as if all the free flowing *tzitzit* in the four corners of the earth could have emerged from this giant source at the center.

There were no people on Main Street when I drove into Lebanon at midday. Only the post office and general store were opened. I went into the post office and asked how Lebanon got its name. As she postmarked stamps with "The Center of USA," the postal clerk said that she had no idea how Lebanon got its name. She sent me across the street to the general store to ask. "You're in luck," the store owner responded. "Every Tuesday and Thursday Gladys Kennedy quilts with her friends at the American Legion hall next door. Gladys knows." He took me next door and introduced me to Gladys as the town historian. As she quilted, she explained that it was named for the cedars of Lebanon that King Solomon used to build the Temple in Jerusalem. She quilted a few more stitches and added, "I have a large cedar growing in my back yard, one of the many growing in this part of Kansas." Gladys walked home to fetch a history of Lebanon for me to get the official version. She returned with a hardbound centennial volume, *A Century at the Center: 1887–1987*. I copied the following from the book while Gladys went back to quilting:

> Name of Lebanon Chosen: A group of early settlers asked Jackson "Jack" Allen, an early settler of the community, to choose a name for the post office and village. Mr. Allen, a Bible student, was the leading literary man of the day and the settlers looked to him to select the name by which the post office town should be designated. "Why not the Bible?" was his first inspiration, and searching the pages, he stopped when he read The Cedars of Lebanon and suggested that name. Nobody opposed, and the name of Lebanon was recorded in the records. This was the year of 1873.

A mile out of town there is an official monument with a bronze plaque marking the geographical center of the United States. It is made of fieldstones stacked into a truncated pyramid holding a flagpole flying the stars and stripes. On the pebbled ground in front of the monument, I drew a left-handed spiral with golden earth from Jerusalem. I framed the Jerusalem spiral by drawing four corners with sand that Miriam and I collected from the four corners of America. Sand from a Florida beach and from between granite boulders on the Maine coast formed the two corners on the right side of the spiral and sand from the beach where San Diego touches Mexico at the Pacific Ocean and from Neah Bay at the tip of the Olympia Peninsula in the State of Washington formed the two corners on the left side of the Jerusalem spiral.

I scooped up some black Kansas soil at America's center and brought it with me to Jerusalem, honored as the Center of the World by both Judaism and Christianity. With the Kansas earth, I drew a right-handed spiral on a slab of stone beside the Western Wall retaining the Temple Mount. This Jerusalem stone combined with cedar wood from Lebanon was used by King Solomon to build the Temple in Jerusalem three millennia ago. As I was photographing the Kansas-earth spiral in Jerusalem to juxtapose with the matching Jerusalem-earth spiral that I photographed in Lebanon, Kansas, I realized that the biblical word "Lebanon" means "heart of the fifty." The first part of the word means "heart" in Hebrew and the second part is the name of the Hebrew letter having the numerical value of fifty. My teenage son, Moshe, enjoyed my playful discovery of Lebanon as the heart of the fifty. Wearing a Miami Panthers T-shirt with *tzitzit* spiraling out from the four corners of his *talit katan* undergarment, he had just returned from having pushed a rolled-up scrap of paper into a space between the huge stones of the Western Wall. Rather than interacting with the rectangular stones, Jews throughout the centuries have related to the open spaces between the stones where they place their hopes and prayers written on small scraps of paper. In our postdigital age, people throughout the world can send their prayers to Jerusalem by e-mail inviting a proxy to print them out, roll them up, and add them to the hundreds of hopes filling the empty spaces between the stones.

Sky art: from Munich to the Tzin Wilderness

My students at the college I headed in the Negev desert helped me tie four *mega-tzitzit* from ship rope and paint one strand skywater blue. We stuffed these 30-foot-long *tzitzit* into four specially made canvas bags to be flown to Germany by Lufthansa. They would hang from the corners of a giant habitable *talit* on the street in front of the BMW Museum in Munich. It would be my art installation for the third international "Sky Art Exhibition."

Since my wife's entire extended family from Holland were murdered by the Germans, I was reluctant to accept an invitation to participate in an exhibition in the city in which Hitler got his start and at a museum across the road from the Olympic Village where 11 Israeli athletes were murdered by Arab terrorists nearly 30 years before 9/11. However, reading the article on Munich in *Encyclopedia Judaica* changed my mind. The enthusiastic support of Munich's citizens for Hitler was no new phenomenon.

> In the second half of the 13th century Munich appears to have had a sizable Jewish community; the Jews lived in their own quarter and possessed a synagogue, ritual bath, and a hospital. On October 12, 1285, in the wake of a blood libel, 180 Jews who had sought refuge in the synagogue were burnt to death.[62]

The anti-Semitic nightmare continued. Munich's Jews were murdered as scapegoats for the plague in 1348, and all the Jews were expelled from Bavaria for the next three centuries in 1442. To harass the Jews during the 18th century, the Munich authorities made it illegal to build a *sukkah*, the traditional hut built for one week each year as a reminder of the Israelites' desert dwellings during their exodus from Egypt. When I looked in my calendar and saw that the opening of "Sky Art '83" fell during the Jewish holiday of *Sukkot*, when each family builds a *sukkah* to celebrate this joyous holiday, I agreed to participate if the City of Munich would support my building a *sukkah* at the entrance to the museum. A *sukkah* is sky art; Jewish tradition requires that stars in the night sky be visible through gaps in its roof. I would design a fringed hut, a giant *talit* sporting four *mega-tzitzit* with blue strands linking sky to sea.

The holiday of *Sukkot* is the culmination of the three biblical pilgrimage festivals in the biblical narrative. *Pesach* (Passover) celebrates the exodus from Egypt, *Shavuot* celebrates receiving the Torah at Mount Sinai, and *Sukkot* celebrates reaching the Promised Land. *Pesach* and *Sukkot* exhibit powerful elements of visual culture that are lacking in *Shavuot* which commemorates the Israelites encounter with the invisible/infinite/eternal author of the Torah. *Pesach* is celebrated by eating *matzah* and participating in an intergenerational performance art event called a *seder*. *Sukkot* is celebrated by holding four species of plants together to symbolize honoring the different personality types that together make up the Jewish people. We also move out of our comfortable houses for one week into fragile huts opened to sky and to our neighbors where we eat and sometimes sleep according to the biblical prescription in *Leviticus 23:39–43*:

> On the fifteenth day of the seventh month, when you harvest the land's grain, you shall celebrate a festival to God for seven days. You shall take for yourselves on the first day the fruit of a citron tree, the frond of a date palm, twigs of myrtle, and brook willows; and you shall rejoice before God for seven days [...] During these seven days you shall live in huts (*sukkot*). Every citizen of Israel shall dwell in huts so that future generations will know that I had the Israelites live in huts when I brought them out of Egypt.

Days before the holiday of *Sukkot*, I arrived at Munich airport. I presented the uniformed German agent with the *menorah* on my Israeli passport and was offered free tourist maps of Munich in a dozen different languages. I chose the Hebrew map. The City of Munich annihilated its Jewish population and then published a map in Hebrew. I never saw Hebrew maps of New York, Los Angeles, or Miami where hundreds of thousands of Jews live today. This Kafkaesque encounter at the Munich airport continued when I was introduced to the city's charming Director of Culture who greeted me in Hebrew. She had learned to speak Hebrew as a volunteer at a kibbutz in Israel where she lived and worked to repent for the sins of her grandfathers. When I arrived at the BMW Museum I found Bavarian pine planks, the same planks used to build the barracks at Dachau death camp, piled on the sidewalk in front of the museum waiting for me to build the *sukkah*. BMW had contributed the wood and sent its carpenters to help me erect the hut. Unfortunately, they refused do anything when they learned that I had no blueprints. It made no difference that I had an accurate drawing of my fringed *sukkah* that I had made for the exhibition catalog. It did not help when I explained that as the designer, I could stand there and direct the construction. "No blueprints! No building!" was their response.

Two other artists overheard my hour-long discussion with the German carpenters and offered to help me build the sukkah. Uri Levi, a systems artist from MIT, and Doron Gazit, an Israeli balloon artist, helped me. As we started to build the sukkah, a Japanese artist passed by and offered to help. Tsutomo Hiroi, Japan's greatest kitemaker who would fly his giant dragons in the Bavarian sky, was the most skilled carpenter of the four of us. He helped us build an elegant and strong structure. As we worked, Hiroi stood inside the sukkah, looked around at it, and chanted, "Ohhh, beautiful Japanese building. Ohhh, beautiful Japanese building." He saw its resemblance to the delicate geometries of rice-paper covered wooden frameworks found in traditional Japanese dwellings. I unsuccessfully tried to convince him that we were building a Jewish building to look like a giant striped prayer shawl. When the sukkah was completed and we hung the mega-tzitzit from the four corners of the structure, he was willing to accept that we had built an Asian building. Israel is on the west coast of Asia while Japan is on its east coast.

The next year, I marked the parentheses of Asia by exchanging sand from the beach in Tel Aviv with sand from the beach at the fishing village of Chikura that I visited with Hiroi. I photographed a parenthesis mark that I etched in the damp beach sand with a stick near the surf line at the Pacific Ocean. I filled the etched arc with yellow Tel Aviv sand. I flew back to Tel Aviv to etch a matching parenthesis mark in the sand at the Mediterranean shore that I filled with black volcanic sand that I had brought to Israel from Chikura. I made a serigraph from the photographs showing the set of two parentheses on stripes of Israel's sky, surf, and sand facing stripes of Japan's sky, surf and sand. The *Parentheses of Asia* serigraph is in the collections of the Emperor of Japan, an oceanographer, and the President of Israel.

When we sat in the *sukkah*, we saw sky between the wooden roof slats that cast shadow stripes on the floor. Jewish tradition requires that the *sukkah* roof, although open to the sky, give more shade than sunlight. The Hebrew word for "shade" *tzel* is related to the word for "salvation" and "rescue" *hatzalah.* The protective shade in the desert provided by the *sukkah* gave the Israelites life-granting refuge from the relentless sun while fleeing from Egyptian bondage. Just as the *sukkah* saved us with its shade, so when we don a *talit* pulling it over our heads, we compare it to divine wings casting a protective shadow on us like the wings of an eagle covering eaglets. *Sukkah* and *talit* are conceptually linked.

We sat and ate in the *sukkah* around a table that I constructed from a clear plastic cylinder holding two discs, one as the tabletop and the second floating midway between the top and the ground. On this second disc, I spread earth flown from Israel to hover over the ground casting an ellipsoid shadow on the *sukkah* floor. My idea for creating a shadow-making table came from my realization that the final two Hebrew letters of *ere**Tz** yisrae**L***, Land of Israel, spell the word for "shadow" ***TzeL.*** Resting in the center of the of disc of earth from the Holy Land was an *etrog*, the beautiful fruit of the citron tree, one of the four species set by the Bible for celebrating *Sukkot*, the holiday called the "Season of Our Rejoicing" in the liturgy.

After the *sukkah* was standing, *tzitzit* attached, and the cylindrical table ready for guests, I rode the tram back to my hotel with several other artists participating in "Sky Art '83." As the tram passed fair grounds with rows of barn-like beer halls (each sponsored by a different beer company), the other artists persuaded me to join them in leaving the tram to experience Munich's Octoberfest. We entered the nearest beer hall. A powerful sudsy aroma hovered over long tables surrounded by blowsy folk in woodsy Bavarian costumes toting enormous steins of beer singing in tune to the um-pa-pa rhythms of a five-piece polka band. As we found seats and were served the sponsor's beer, a new tune began and the entire crowd began to sing out loudly in cadenced unison simultaneously raising their beer steins up high. It looked like a movie set for a period film. The period image that came to mind in horror was my childhood memory of newsreel films of vast crowds raising their arms high together shouting out as one, "Heil Hitler!" I could see Munich's citizens cheering Hitler as he proclaimed the Nazi revolution during his "Beer Hall Putsch." This merging of individuals into an overwhelming oneness that submerges individuality was an altogether different togetherness than I had just experienced building the *sukkah* with Horoi, Uri and Doron.

I closed my eyes and saw the plaza before the Western Wall in Jerusalem. Hundreds of people are praying there at all times of the day. They do not converge at any point to chant their prayers together, as an army of worshipers might do. There are no fixed times for services where everyone could join together in one large assembly. Instead, Jews form ad hoc *minyanim* (prayer quorums). As soon as ten men find themselves together, they begin the prayer service as a few others join them. Dozens of services, each beginning spontaneously can be seen simultaneously. People float in and out of the scene coming together in small groups of strangers who are suddenly spiritually linked for half an hour or so. They never find themselves submerged in an overwhelming oneness that diminishes individual expression.

Marking the opening of the "Sky Art '83" exhibition, an international sky art conference was held at which I was invited to deliver the keynote address. My talk, "Higher than Sky," revolved around a Hassidic tale in which Hassidim tell about their great rebbe who ascends to heaven during the ten days between the high holidays of Rosh Hashanah and Yom Kippur. A skeptic comes to their town and hears them lovingly tell about how their rebbe ascends to heaven in order to plead for the forgiveness for all humanity's transgressions in a face-to-face encounter with God. The skeptic confronted a group of the Hassidim: "How can you think such ridiculous nonsense? According to tradition, even Moses fell short of such a face-to-face encounter." They responded, "If you knew our rebbe, you too would recognize his greatness." One morning in synagogue, the skeptic sees the rebbe, who was seated in the front next to the ark, suddenly disappear. He ran out of the synagogue and spied the rebbe rapidly walking down the street. The skeptic discreetly trailed the rebbe and saw him enter his home to emerge a short time later dressed in workman's clothes with an ax in his belt and a rope draped over his shoulder. The rebbe walked to the edge of his village where the forest began, chopped down a small tree, cut off its branches, tied all the wood together with his rope, and entered a shack with the bundle of wood on his back. Peering through a window, the skeptic saw a frail old woman in bed and the rebbe putting the wood in her stove, peeling potatoes, and putting up a stew to cook, changing her bedding, and getting down on his knees to scrub the floor. He then spied the rebbe walking back home, replacing his work clothes with an elegant black brocade robe and a white woolen *talit*, and returning to the synagogue through a back door. The skeptic quietly slipped into the synagogue to find the Hasidim talking ecstatically about their rebbe's return from his ascent to heaven. The skeptic added, "If not higher than that!"

The skywater blue strand of the *tzitzit* flowing from the corners of a *talit* symbolizes sky flowing down to earth as a reminder that acts of kindness are the highest expression of human values. Being down on one's knees scrubbing the floor for an old invalid woman is the way to reach higher than sky. Moreover, the *sukkah* symbolizes all human beings living in peace with each other while celebrating the holiday of *Sukkot*, the "Season of Our Rejoicing." All people were invited into our Munich *sukkah* to share our joy. This invitation follows from the biblical invitation to all nations of the world from *Zechariah 14:16–19*, which is read in synagogues on *Sukkot*. The prophet Zechariah teaches that if all the people of the world would live for just one week in fragile huts open to their neighbors and to the sky, then peace with each other and nature would follow, thereby ushering in the Messianic Age.

The weeklong holiday of *Sukkot* ended with a star-filled Bavarian sky. As my sky art event, I had planned to release five-foot-high Styrofoam Hebrew letters into the sky lifted by helium-filled weather balloons. Searchlights would illuminate them as they ascended over Munich. This visual *midrash* is based upon a *midrash* that relates to the seven Hebrew letters in the Torah scroll that are written by the scribe with little three-pronged crowns on them called *tagin*. They are letters in heavy words like *sinah* "hate" that are too heavy to ascend to heaven when the Torah text is chanted. The *tagin* provide extra lift heavenward to letters weighed down by their connection to conflict. I painted each letter one of the seven colors of the rainbow and attached three balloons to each one as giant *tagin*. I consulted with the Bavarian meteorological services to determine the size of weather balloons that would lift the letters into the jet stream so that they would fly eastward into the Soviet Union where the Iron Curtain was slammed shut on Jews who wished to escape from anti-Semitic harassment. I enthusiastically envisioned MIGs scrambling to intercept Hebrew letters invading Soviet airspace.

However, it didn't happen. As I was leaving the hotel that night, American artist Lowry Burgess, creator of the first art satellite placed in orbit by NASA, intercepted me looking distraught. He was holding a steel-gray plastic bag in one hand and a smashed *etrog* cradled in his other hand. In a distressed voice, he told me how a neo-Nazi motorcycle gang had attacked my *sukkah*. They tried to destroy the *sukkah* with crowbars and steel chains. Thanks to Hiroi's help, the *sukkah* was strong enough to survive their blows. However, they succeeded in destroying the table, smashing the *etrog* and scattering the earth over the ground. They tied hangmen's nooses in the rope of the *tzitzit*. Lowry said, "I didn't think that you would want to have holy land thrown out in the garbage in Germany. So, I swept it up for you and put it in this plastic bag." Realizing that Hebrew letters could not fly free in Germany, I cancelled the event. Instead, I descended into the depths of the earth with the letters. The seven Hebrew letters rode the escalators between rush-hour commuters at the subway stop shared by the BMW Museum and the Olympic Village where Arab terrorists murdered eleven Israeli athletes in cold blood.

The next morning, Uri Levi and I took the commuter train from Munich to the suburbs. I carried the bag of earth. We exited the train under the large sign: DACHAU. It was an ominous experience for two Jews. Walking down from the raised station into the center of a shockingly beautiful town gave Uri and me a bout of aesthetic flu. In the mist of this floral suburb with every blade of grass and tree trimmed, every pastry displayed in exquisite taste in the shop window, every house freshly painted clean white, Hitler built his first death camp. Middle-class Germans lived a middle-class life in their garden paradise while the cries of thousands of Jews being tortured and brutally murdered in their midst went unheard. I had erroneously thought that there was some connection between aesthetic and moral development of human beings.

We walked from the Dachau train station to the rebuilt death camp taking turns carrying the bag of earth from the Land of Israel. Allied bombers had destroyed the original. A true to scale, neat, trim reproduction of the former death camp was rebuilt out of the same lovely Bavarian pine planks that BMW supplied for my *sukkah*. At the foot of a concrete pillar supporting the barbed-wire fence surrounding the camp, I spread the swept remains of the scattered earth from the Holy Land. The earth rested on freshly mowed grass that covered up bloodied ground. I was following an ancient Jewish tradition of placing earth from the Land of Israel in the graves of our dead in the lands of our exile. On a square of earth from my sky art *sukkah* spread out on the grass, I set steel rebar rods that I had found discarded at a construction site on my walk from the Dachau town center. With the rods, I wrote out the word *sukkah* in three square Hebrew letters, the first letter is totally closed, the second is open on one side, and the third is open in two places. The form of the letters in the word *sukkah* can be metaphorically read as "towards freedom." Above and below the Hebrew word *sukkah*, I wrote the word *sukkah* in rebar rods two more times upside-down and backwards. The German iron cross and swastika were trapped between the nine letters.

Uri dropped to the ground and wept. I paced furiously to express my anger.

It was intolerable for me to look at the photographs that I took of my earth art memorial on the verdant grass of Dachau with lovely bushes growing up against a bright blue sky. They failed to give any indication of the horror of the place. After years of not showing these photographs, I realized that I could transform a sunny day into a dark day in hell by removing my slide from its frame and printing it as if it were a negative. Printing the positive slide resulted in a negative image in which bushes become rising flames and sky and grass different shades of deadly brown.

After the Sky Art show opened, Lowry Burgess returned to Israel with me. He had collected water from the major rivers of the world, the Nile, the Amazon, the Mississippi, the Yangtze, etc. We went from my home in Yeroham in the Negev desert mountains to Beersheva where we bought chemical glassware to build a distillation apparatus. We then drove down to the Dead Sea where we assembled the distillation apparatus on a salt encrusted rock in the Sea. Lowry mixed the river waters together and we distilled the mixture. It was as if the waters of all the major rivers of the world flowed down to the lowest spot on planet Earth. Lowry set this distillate in the core of the satellite that NASA placed in orbit as part of his *Quiet Axis*, a narrative artwork that was decades in the making. *Quiet Axis* reveals his ecological perspective as he links the satellite orbit with an axis that he began creating to extend from the Bamiyan desert in Afghanistan to the other side of the planet beneath the Pacific Ocean near Easter Island.[63]

The rainbow of seven human-size Styrofoam Hebrew letters that were slated to announce themselves in the Bavarian sky and pierce the Iron Curtain could not fly free on the European continent drenched in Jewish blood. They would fly free in the Tzin wilderness separated from the Dead Sea by the Negev desert that drops down to the lowest place on the planet through two colossal craters. This was the entry point into the Promised Land taken by the spiritual leaders of the twelve Israelite tribes to spy out the land. "The men headed north and explored the land from the Tzin Wilderness all the way to Reẖov" (*Numbers 13:21*). At the edge of a rocky cliff overlooking the Tzin wilderness, my art students worked with me to tie weather balloon *tagin* on the tops of each of the letters. The large red balloons were filled from a tank of hydrogen. Helium, only made in United States, was unavailable. We tethered the letters to rocks and planned to release them simultaneously. Unexpectedly, before we were ready to release the letters, a sudden gust of wind ripped the letter *zayin* loose, setting it free. As it ascended over the Tzin wilderness, an eagle spiraled around it escorting it up into a cloud.

Notes

1. Irit Rogoff, "Studying Visual Culture," in Nicholas Mirzoeff (ed.), *The Visual Culture Reader* (London and New York: Routledge, 2002), p. 24.
2. Suzi Gablik, *The Reenchantment of Ar t* (New York: Thames and Hudson, 1991), p. 7, 8.
3. Jean Gebser, "The Foundations of the Aperspective World," *Main Currents in Modern Thought*, 1972, vol. 29, pp. 80–88, and *The Ever-Present Origin*, trans. Noel Barstad and Algis Mickunas (Athens: Ohio University Press, 1985).
4. Algis Mickunas, "Civilizations as Structures of Consciousness," *Main Currents in Modern Thought*, vol. 29, 1973, pp. 179–185, and Eric Mark Kramer and Algis Mickunas, "Gebser's New Understanding", in Eric Mark Kramer (ed.), *Consciousness and Culture* (Westport, CT: Greenwood Press, 1992), pp. x–xxxi.
5. Gablik, *The Reenchantment of Art.*
6. Julian Jaynes, *The Origin of Consciousness in the Breakdown of the Bicameral Mind* (Boston: Houghton Mifflin, 1976).
7. Eric Mark Kramer (ed.), *Consciousness and Culture* (Westport, CT: Greenwood Press, 1992), p. xvii, original emphasis.
8. Justin Schorr, *Toward the Transformation of Art* (Rutherford, NJ: Fairleigh Dickinson University Press and London: Associated University Presses, 1974).
9. George T. Lock Land, *Grow or Die: The Unifying Principle of Transformation* (New York: Delta, 1973).
10. Thomas Kuhn, *The Structure of Scientific Revolutions* (Chicago:University of Chicago Press, 1962).
11. Edwin A. Abbot, *Flatland*, unabridged edition (New York: Dover, 1992 [1884]).
12. Stephanie Barron, *"Degenerate Art": The Fate of the Avant-Garde in Nazi Germany* (Los Angeles: Los Angeles County Museum of Art, 1991).

13. Pinchas H. Peli, *The Jewish Sabbath* (New York: Schocken, 1988), p. 102, 103.
14. Edmund Carpenter, "Image Making in Arctic Art," in Gyorgy Kepes (ed.), *Sign, Image, Symbol* (New York: George Braziller, 1966).
15. Michael Cole and Sylvia Scribner, *Culture and Thought* (New York: Wiley, 1974).
16. W. Hudson, "Pictorial Depth in Sub-Cultural Groups in Africa," *Journal of Social Psychology*, 52:1, 1960, and "The Study of the Problem of Pictorial Perception Among Unacculturated Groups," *International Journal of Psychology*, 2:1, 1967.
17. J. B. Deregowski, "Difficulties in Pictorial Depth Perception in Africa," *British Journal of Psychology*, 59:3, 1968, and with W. Blyth, "Hudson's Pictures in Pandora's Box," *Journal of Cross-Cultural Psychology*, 1:4, 1970.
18. Arnold Hauser, *The Social History of Art* (New York: Random House, 1951).
19. P. A. Michelis, "Byzantine Art as a Religious and Didactic Art," *British Journal of Aesthetics*, 7:2, 1967.
20. Louis Rapkine, "Notes for a Scientific Theory of Aesthetics," *Leonardo*, 3:4, 1970.
21. M. C. Goodall, *Science and the Politician* (Cambridge, MA: Schenkman, 1965).
22. Paul B. Weisz, *The Science of Biology* (New York: McGraw-Hill, 1963).
23. Arthur David Richie, *Studies in the History and Methods of the Sciences* (Edinburgh: The University Press, 1958).
24. Al Gore, *Earth in Balance: Ecology and the Human Spirit* (New York: Plume, 1993).
25. Jurgen Claus, "Terminal Art," in Timothy Druckery (ed.), *Ars Electronica: Facing the Future* (Cambridge, MA: MIT Press, 1999), p. 373.
26. E. Lozano and A. Mickunas, "Gebser and Pedagogy: The Integral Difference," in E. M. Kramer (ed.), *Consciousness and Culture* (Westport, CT: Greenwood, 1992), pp. 179–199.
27. Mel Alexenberg, "Toward an Integral Structure Through Science and Art," *Main Currents in Modern Thought,* 30: 4, 1974; and Mel Alexenberg, "Bridging Cultures Through Interdisciplinary Art Education," in Mary Stokrocki (ed.), *Interdisciplinary Art Education* (Reston, VA: National Art Education Association, 2004).
28. Mel Alexenberg, *Aesthetic Experience in Creative Process* (Ramat Gan, Israel: Bar-Ilan University Press, 1981).
29. David Bohm, *Wholeness and the Implicate Order* (London: Routledge & Kegan Paul, 1980).
30. Michael Talbot, *The Holographic Universe* (New York: HarperCollins, 1991), p. 46.
31. Karl Pribram, *Languages of the Brain* (Monterey, CA: Wadsworth, 1977).
32. Paul Pietsch, *Shufflebrain: The Quest for the Holographic Mind* (Boston: Houghton Mifflin, 1981).
33. Charles Darwin, *The Origin of the Species* (New York: Wiley 1955).
34. H. W. Jansen, *History of Art* (New York: Abrams, 1991).
35. Rosanna Vitale, "The Significance of Aperspectival Art in Light of Gebser's Work," in Eric Mark Kramer (ed.), *Consciousness and Culture* (Westport, CT: Greenwood Press, 1992), p. 111.
36. Claude Cernuschi, *Jackson Pollock: Meaning and Significance* (New York: HarperCollins, 1992).
37. Ibid.
38. Allan Kaprow, "The Legacy of Jackson Pollock," *ARTnews*, vol. 57, October 1958.
39. Allan Kaprow, "The Happenings are Dead…Long Live the Happenings," *Artforum*, vol. 4, March 1966.
40. Cernuschi, *Jackson Pollock: Meaning and Significance.*
41. Robert Smithson, "The Spiral Jetty" in Nancy Holt (ed.), *The Writings of Robert Smithson* (New York: New York University Press, 1979).

42. Ori Z. Soltes, *Jewish Artists: On the Edge* (Santa Fe, NM: College of Santa Fe, 2000), p. 28.
43. Donald Kuspit, "Is the Concept of Jewish-American Art Viable?" in *Jewish Artists: On the Edge*, p. 11.
44. Ibid.
45. I. Hassan, "The Culture of Postmodernism," *Theory, Culture and Society*, 2: 3, 1985, p. 123, 124.
46. David Harvey, *The Condition of Postmodernity* (Cambridge, MA and Oxford, UK: Blackwell, 1990), p. 116, 42.
47. Ibid., p. 51.
48. Lowry Burgess, "Body and Garden in Zero State: The Disembodiment of the Embodied; The Embodiment of Disembodied," address to the International Dance and Technology Conference, Arizona State University, Tempe,1999, www.art.cfa.cmu.edu/burgess/essay.html.
49. Ilya Prigogine and Isabelle Stengers, *Order Out of Chaos: Man's New Dialogue with Nature*, p. 12.
50. Doron Witzum, Eliyahu Rips, and Yoav Rosenberg, "Equadistant Letter Sequences in the Book of Genesis," *Statistical Science*, 1994, 9: 3, pp. 429–438. For a general discussion of computer codes in the Bible, see Michael Drosnin, *The Bible Code* (New York, Simon and Schuster, 1997) and Moshe Katz, *Computorah* (Jerusalem: Achdut, 1996).
51. Walter Mopps and Susan Davidson, *Robert Rauschenberg: A Retrospective* (New York: Guggenheim Museum, 1997), pp. 56–59.
52. Ibid., p. 552. For a discussion of John Cage's collaboration with Robert Rauschenberg and Merce Cunningham as a seminal figure in the transition from modernism to postmodernism see: Nancy Perloff, "'The right to be myself, as long as I live! As if I were a sound': Postmodernism and the Music of John Cage," in Hans Bertens and Joseph Natoli (eds.), *Postmodernism: The Key Figures* (London: Blackwell, 2002).
53. Aryeh Kaplan, *Sefer Yetzirah: The Book of Creation* (York Beach, ME: Samuel Weiser, 1993).
54. D'Arcy Thompson, *On Growth and Form* (London: Cambridge University Press, 1917; 1942). Abridged and edited by John Tyler Bonner (London: Cambridge University Press, 1961).
55. Delta Willis, *The Sand Dollar and the Slide Rule: Drawing Blueprints from Nature* (Reading, MA: Addison-Wesley, 1995).
56. *Babylonian Talmud: Chullin, 89.*
57. *Babylonian Talmud: Menachot, 44a.* Based upon narration in Aryeh Kaplan, *Tzitzith: A Thread of Light* (New York: NCSY/Orthodox Union, 1984), p. 49.
58. Menachem M. Schneerson, *Torah Studies*, adapted by Jonathan Sacks (London: Lubavitch Foundation, 1986), p. 242.
59. Gideon Weitzman, *Sparks of Light: Essays on the Weekly Torah Portions Based on the Philosophy of Rav Kook* (Northvale, NJ, and Jerusalem: Jason Aronson, 1999), p. 207.
60. Tom Anderson, "Toward a postmodern approach to art education," in J. Hutchins and M. Suggs (eds.), *Art Education: Content and Practices in a Postmodern Era* (Reston, VA: National Art Education Association, 1997), pp. 62–73.
61. Rashi, Commentary on *Bamidbar 15:41.*
62. Munich, *Encyclopedia Judaica*, vol. 12, p. 522.
63. Lowry Burgess, *Burgess: l'axe tranquille=Burgess: el eje tanquilo=Burgess: the quiet axis* (Montreal: Editions du Trecarre, 1987).

Chapter 4

Kabbalistic Perspectives: *Creative Process in Art and Science*

Whoever is endowed with the soul of a creator must create works of imagination and thought, for the flame of the soul rises by itself and one cannot impede it on its course [...] The creative individual brings vital, new light from the higher source where originality emanates to the place where it has not previously been manifest, from the place that "no bird of prey knows, nor has the falcon's eye seen." (Job 28:7), "that no man has passed, nor has any person dwelt" (Jeremiah 2:6).[1]

Abraham Yitzhak Kook

If a man wishes to attain the rank of holiness, he must become a creator of worlds. If a man never creates, never brings into being anything new, anything original, then he cannot be holy unto his God. That passive type who is derelict in fulfilling his task of creation cannot become holy. Creation is the lowering of transcendence into the midst of our turbid, coarse, material world [...] "For the Lord thy God walketh in the midst of thy camp" (Deuteronomy 23:15).[2]

Joseph B. Soloveitchik

"In the beginning God created *et* the heaven and *et* the earth" (*Genesis 1:1*). These are the first words of the Bible. In the original Hebrew, *et* is the first creation, the blueprint for creating heaven and earth. In the English translation, the word *et* drops out since it has no English equivalent. The word *et* is a grammatical form indicating a direct object linking verb and noun. It links "God created" to "heaven" and to "earth." It is spelled *alef-tav*, the first and last letters of the Hebrew alphabet. Spanning the full set of 22 Hebrew letters from *alef* through *tav*, *et* represents media systems. "Heaven" represents spiritual systems and "earth" natural systems.

Postdigital age thought gives us tools to comprehend relationships between the three divine creations in the first biblical passage: media, spiritual, and natural systems. Hebrew letters and words are the basic elements of the biblical media system called *Torah*. The 22 Hebrew letters are the pathways between ten gateways (*sephirot*) in spiritual systems revealed through *kabbalah*. Atoms and molecules are basic elements of natural systems. And bits and bytes are the basic elements of digital media systems. Science reveals the secrets of natural systems that have led to decoding genetic media systems and to the development of digital media systems. *Kabbalah* reveals the secrets of spiritual systems by decoding the biblical media system to teach ways of living as compassionate creators. Art's role serves an integrative function. The challenge facing the postdigital artist is to make art that brings material, spiritual, and media worlds together in creative interplay.

Spiritual bits and bytes

"God said, 'Let us make a human being in our image, after our likeness." (*Genesis 1:26*). We actualize being in the divine image and likeness by sharing the ability to create new worlds with God. The biblical text reads "Let us make," rather than "Let Me make." It teaches that we are partners of God in our own self-creation. The Hebrew word for "image," *TZeLem* is the core of the name of the biblical artist *BeTZaLel*, which can be translated "In God's Image." Betzalel is said to have had the divine secret of forging combinations of the 22 Hebrew letters to create new worlds.

The postdigital era makes this kabbalistic notion of artistic creativity through making permutations of bits of information more than a quaint legend. It is computer science rather than mysticism, physics rather than metaphysics that lets us reveal in our times this ancient wisdom. All the multitude of words, sounds and images that we can access today from the Internet, CDs, DVDs, are encoded in bits strung together in groupings of eight bits called bytes. The 256 bit permutations in one byte are in turn grouped into billions of combinations that we perceive as a website, a blog, a computer game, a text, a song, or a movie.

Hebrew letters and words have numerical values in the decimal system like electronic bits and bytes in the binary system. *Gematria* is a system for exploring mathematical relationships between Hebrew words to find spiritual significance.

> The twenty-two sacred Hebrew letters are profound, primal, spiritual forces. They are in effect, the raw material of Creation [...] The letters can be arranged in countless combinations, by changing their order within words and interchanging letters in line with the rules of various Kabbalistic letter systems. Each rearrangement results in a new blend of cosmic spiritual forces.[3]

Rabbi Nosson Scherman, who wrote these words as an introductory overview to the Hebrew prayer book, *ArtScroll Siddur*, continues with an analogy from the physical realm of the combinations of atoms into molecules. The different numbers of protons, neutrons, and electrons that make up the atoms of each of the 92 different elements determine its properties. These atoms, in turn, combine into molecules, and molecules into supersized molecules like DNA in which all of life's forms are written in a binary language of two pairs of two letters: A-T, T-A, and C-G, G-C. All computer information is written in a binary system of 0 and 1, like a circle and a line. The kabbalistic model called the "Tree of Life" is constructed from ten circles, the *sephirot*, interconnected by 22 lines, representing the Hebrew letters. It provides a symbolic language and conceptual schema for exploring two parallel creative processes – human and divine. The description of Betzalel's qualities as an artist and *kabbalah* are derived from the same biblical passage (*Exodus 31:2*). "Wisdom, understanding, and knowledge" are both the artist's cognitive traits and the highest of the ten *sephirot* in the kabbalistic model. The *sephirot* are stages in the creative process that bring divine light down into our everyday world.

Just as each Hebrew letter has a numerical equivalent, kabbalists have assigned colors to each letter. Using a color scheme based upon the rainbow spectrum, I created the interactive artwork *Torah Spectrograph* at MIT. Through computer graphics, it reveals hidden patterns in the Torah. The participant enters his birthday into the computer from which it calculates his birthday according to the Hebrew calendar. He can then watch his bar mitzvah portion scrolling down the monitor in bands of color. A digital version of the first five books of the Bible in the original Hebrew was combined with a look-up table programmed to display colored bands for each letter. The seven colors of the spectrum were repeated three times corresponding to 21 of the 22 Hebrew letters. The first letter, *alef*, represents the number one and unity. It unifies all the colors into white light and is displayed as a single white pixel. The second letter, *bet*, is two red pixels, the third letter, *gimel*, is three orange pixels, until we come to the final letter, *tav*, which is a violet band 22 pixels long. Each of the 56 Torah portions exhibits a unique set of patterns and color relationships.[4]

Qualities resulting from quantity and arrangement of modules in both the spiritual realm of Hebrew letters and the physical realm of atomic structure find a parallel in the realm of digital media. All computer information is encoded in bits, the smallest elements of information. In *Being Digital*, MIT Media Lab Director Nicholas Negroponte explains: "A bit has no color, size, or weight, and it can travel at the speed of light [...] It is a state of being: on or off, true or false, up or down, in or out, black or white. For practical purposes we consider it to be a 1 or a 0."[5] He tells about his visit to a top high tech company where he was asked the value of his laptop computer at the sign-in desk. Negroponte responded that it was worth between one and two million dollars. The receptionist insisted that it could not be worth more than $2,000 and wrote down that amount. "The point is that while the atoms were not worth that much, the bits were almost priceless."[6] It is with these bits that Negroponte created MIT's Media Lab, *Wired* magazine, and his seminal writings. In the digital age, Microsoft, which is in the business of selling immaterial bits, has four times the revenues as America's top steel maker, US Steel, which sells material atoms.

In *The Language of New Media*, Lev Manovich identifies the first principles of new media as numerical representation and modularity, both central to the kabbalistic paradigm.

> *Numerical Representation.* All new media objects, whether created from scratch on computers or converted from analog media sources, are composed of digital code; they are numerical representations. *Modularity.* This principle can be called the "fractal structure of new media." Just as a fractal has the same structure on different scales, a new media object has the same modular structure throughout. Media elements, be they images, sounds, shapes, or behaviors, are represented as collections of discrete samples (pixels, polygons, voxels, characters, scripts). These elements are assembled into larger scale objects but continue to maintain their separate identities.[7]

The principle of modularity and fractal structures parallels the kabbalistic concept of interinclusion in which all ten *sephirot* are included in each *sephirah* (singular form of *sephirot*) like the branching patterns of a tree repeated in the venation patter ns of its leaves. Interinclusion mirrors the holographic concept that all the information is contained in each part of a hologram. In a hologram cut apart with a scissors, the whole picture can be played out with laser light from each part, unlike a photographic image cut into pieces. Understanding new technologies helps us grasp the ancient wisdom of how the 32 information elements (10 *sephirot* and 22 letters) in the kabbalistic paradigm interrelate in myriad ways in the process of creating new worlds.

Sepher Yetzirah: The Book of Creation,[8] is considered the oldest and most mysterious of all kabbalistic texts. "With 32 mystical paths of Wisdom," the book begins, "the Universe was created with three *sepharim*, with *sepher* (form), with *s'phar* (quantity), and with *sipur* (narrative)." We have discussed earlier the *SPR* root of *SePheR Torah*, the Torah scroll, and the significance of the *SPiRal* scroll form in conveying the message of endless flow, of inSPiRation and SPiRituality. In *The Book of Creation*, *SePheR* refers to the form of each Hebrew letter and the meaning the form conveys. The first letter of the Torah is *bet*, which is the prefix "in" as "In the beginning" (*Genesis 1:1*). Its form is a square open on the left side, like a house with an open door. The top horizontal line of *bet* symbolizes heaven, the bottom line earth, and the right vertical line draws heaven down to earth. The entire Torah flows out from the open side. As mentioned earlier, the negative space around the letter *bet* forms the letter *pei*, which means mouth, a metaphor for the divine mouth uttering the words of the Torah.

S'PhaR is the numerical value of each letter (related to the word "cipher"). Using the *gematria* system, we can calculate the numerical values of the biblical Hebrew words to discern spiritual messages from quantitative relationships between words. Since the Hebrew words for "ladder," "spiral," and "Sinai" all equal 130, we learn that the ladder in Jacob's dream was a spiral ladder like DNA that symbolizes receiving the Torah at Sinai brought down from the clouds to earth. *SiPuR* means story, letters strung together in words and words sequenced in the biblical narrative of creation, revelation, and redemption.

Parallel to *kabbalah*'s locating number, form, and process in the spiritual realm, is the significance of number, form, and process in science's discoveries in natural systems. Number (*s'phar*) determines the qualities of matter. The smallest atom with one electron circling one proton is manifest as the gas hydrogen. Atoms that have four electrons and four protons are carbon, from coal to diamonds. Six are oxygen, 26 iron, 79 gold, and 92 uranium. At the molecular level, carbon atoms surrounded by hydrogen atoms strung together like words in a sentence form the paraffin series (C_nH_{2n+2}). One carbon, two carbons, and three carbon chains are cooking gas. Chains with eight are liquid gasoline, and long chains of 20 carbons are solid candle wax. If carbon chains become rings, where one end is connected to the other like a necklace, a totally changed material emerges. A different form (*sepher*) with the same number of carbon atoms creates a material qualitatively different.

The most important story (*sipur*) of life on this planet is a process also involving carbon and hydrogen atoms called photosynthesis ($6CO_2+6H_2O \longrightarrow C_6H_{12}O_6+6O_2$). Molecules of carbon atoms linked to two oxygen atoms enter the leaves of green plants and molecules of oxygen atoms linked to two hydrogen atoms flow up into the plant from its roots. In the leaves, with the aid of chlorophyll and sunlight, carbon dioxide and water are transformed into glucose and leftover oxygen is released into the air. Without this process, we would have neither food to eat nor oxygen to breath. There would be no life on planet Earth.

The number and sequence of four compounds as the rungs of DNA double helix ladders determines a substantial part of our physical and mental traits. Human beings have 23 pairs of DNA ladders in every cell of our body. Having three instead of two of the 21st chromosome is expressed as Down's syndrome. Having 16 pairs of DNA ladders rather than 23 would make you an onion rather than a human being. The information for all forms of life from the amoeba to the giant redwoods, from roses to elephants, from bacteria to whales, and from onions to human beings, is written with an alphabet of only four letters like the biblical Tetragrammaton. They are A for adenine, T for thymine, C for cytosine, and G for guanine. These letters spell four words: A-T, T-A, C-G, G-C. Each rung of the DNA is written with one of these words. The sequence and number of these words determines whether you have blue eyes or brown, whether you will be short or tall, and whether you will be a genius or mentally retarded. Since all forms of life are written with the same four words, genetic engineering has developed the ability to transpose a DNA sequence from one organism to another to create a new transgenic organism. The sequence in lilacs that made them purple can modify the DNA of a tomato to create a purple tomato. The DNA sequence that makes jellyfish glow green in the dark can give other animals or plants that same ability.

The transgenic art of Eduardo Kac interrelates biblical, digital, and genetic media systems. Kac coined the term "transgenic art" as art that appropriates research findings in genetics and cell biology to create a new class of transgenic life forms. His work explores the metaphorical implications of presenting transgenic organisms as art. His most well-known transgenic organism is Alba, a white rabbit into which the gene from a jellyfish that codes for the production of green florescent protein (GFP) was introduced. The modified DNA code in the rabbit is expressed as bioluminescence. Alba glows green like the jellyfish in the sea from which the GFP genetic sequence was taken. Kac expresses the biblical concept of the artist as a partner of God in continuing the Creation in presenting Alba as an artwork titled *GFP Bunny*. He further develops this biblical concept in his transgenic artworks *Genesis* and *The Eighth Day*.

Genesis is an artwork in which a genetic media system was informed by the biblical media system and an early form of digital media system.

> Kac created a synthetic gene by transposing a sentence from the biblical book of *Genesis* into Morse Code, then translating the code into DNA base pairs according to a translation scheme devised by the artist [...] This 'Genesis gene was incorporated into bacteria, which formed part of a display. Remote participants using the Web could switch an ultraviolet light in the gallery on and off, causing biological mutations in the bacteria. When the base pairs of the mutated bacteria were decoded, the meaning of the original biblical text was subtly transformed.[9]

Eduardo Kac created his transgenic artwork *The Eighth Day* as a manifestation of the Torah view that the God's seven days of creation is followed by the eighth day when people are invited to become God's partners in continuing the process of creation. Art historian Edward Lucie-Smith shows how his earlier work, *Time Capsule*, is an expression of Kac's Jewish consciousness. The artist inserted under the skin of his ankle a tiny microchip containing a programmed identification number sealed in biocompatible glass. This is a common procedure for identifying lost dogs. When scanned, the implant generates a low frequency radio signal causing the microchip to transmit its numerical code. Kac questions the legitimacy and ethical implications of such procedures in a postdigital culture.

> In fact, in his case the work has a very precise meaning, since a number of his family were Polish Jews who perished in the Nazi Holocaust. Photographs of some of these lost relatives were included as part of the display in the gallery where the insertion of the microchip took place. The microchip incorporating a number alludes to the numbers tattooed on the arm of those who were herded into concentration camps. It also alludes to the fact that the genetic heritage of those lost relatives lives on in Kac's own body.[10]

Kac created *The Eighth Day* at the Institute for Studies in the Arts at Arizona State University in collaboration with researchers from the Department of Biology. He created a contained ecosystem that brought together living transgenic life forms and a biological robot (biobot) under a clear four-foot-diameter acrylic dome. The organisms that inhabited his ecosystem were mice, zebra fish, and tobacco plants, all genetically modified to emit green light. These bioluminescent life forms shared their environment with a biological robot containing a colony of amoebae also genetically modified to emit green light. A bioreactor monitored changes in amoebae activities that caused the biobot to move up and down inside the transparent dome. In contrast with the Hellenistic single-point perspective of passive observer, Kac designed his biobot to provide multiple perspectives from both inside and outside the system as a realization of Hebraic consciousness. The visitors to the gallery were able to peer into the terrarium from the outside and also place themselves inside the system through the viewpoint of the biobot as if they were living in the dome. Visitors who saw through the biobot's "eyes" were joined by participants worldwide who could look into the bioluminescent ecosystem via a website that Kac created. They could experience living in a world cohabited by glowing creatures.

Biblical roots of *kabbalah*

Postmodern art is about interactive systems rather than static objects. It honors process more than product. The prototype for this postmodern dynamic is the process of art-making itself, the creative process that transforms mental worlds of thoughts and feelings into environmental expressions. The description of this process derived from the deep structure of biblical consciousness is *kabbalah*. It originates in the account of the creation of the Tabernacle under the artistic direction of Betzalel "See, I have called by name: *Betzalel ben Uri ben Hur*, of the tribe of Judah. I have filled him with a divine spirit, with Wisdom, Understanding, and Knowledge, and with the talent for all types of craftsmanship" (*Exodus 31:2*). That his name is "In the Divine Shadow" rather than "In the Divine Light" acknowledges the shadow side of the creative process, the Freudian subconscious, the dark inclinations that need to be transformed into life-enhancing energies. The artist possesses the creative power to turn darkness into light.

Art and *kabbalah* originate in the same biblical passage. The artist's traits "Wisdom, Understanding and Knowledge" are the first stages in the kabbalistic model of human creativity that parallels divine creation. *Kabbalah* is Judaism's esoteric tradition that provides a symbolic language and conceptual schema for exploring the creative processes. The KBL root of the Hebrew word *KaBaLah* appears in the Bible itself in the word *maKBiL* meaning, "parallel."[11] The artists creating the Tabernacle covered it with two large tapestries each having 50 loops parallel to each other, linked together with gold fasteners (*Exodus 26:5* and *Exodus 36:12*). The tapestries symbolize divine creation and human creativity linked one to the other. Since these two creative processes are parallel, we can learn about the Creation of the universe through gaining insight into our own creative process.

Kabbalah literally means "to receive." It is the hidden wisdom of the deep structure of biblical consciousness received from generation to generation. I hear the word *kabbalah* spoken frequently in Israel where I live. I hear it from the supermarket checkout clerk when she hands me the long paper ribbon saying, "*kabbalah shelkhah*," "your receipt." It is appropriate and significant that both a supermarket computer printout and the Jewish mystical tradition share the same word. I appreciate the esoteric supermarket language – the bar code on boxes, bottles and cans. We all stand illiterate before the secret language of the digital age that only supermarket lasers can read. *Kabbalah* is the secret language of down-to-earth mysticism that explores how divine energies are drawn down into our everyday world.

Studying *kabbalah* invites the learner to visualize its symbolic language in terms of concrete experiences. The deepest structure of biblical consciousness can only be understood at the level of everyday life. Rabbi Zalman Schachter tells a Hasidic story to illuminate this core concept in understanding the kabbalistic model of creative process.

Shmuel Munkes was walking down a road on his way to see the rebbe when an elegant carriage stops. A well-dressed dandy invites him to ride with him since he is going to see the rebbe, too. The dandy brags about being the son and grandson of kabbalists. Shmuel asks this self-proclaimed kabbalist for help in deciphering a kabbalistic text of cosmic proportions that he said he had found on a scrap of paper in a holy book:

> In the very primal beginning there was chaos – all was sundered and separate. Grainy nuclei unconnected. Swirling. Then fiat, they were one in one sphere. The sphere unfolded into an orb. On the orb-lines appeared, forces cut the space in fields. These fields became

> centered in a point and enfolded the point. Peace was made between fiery angels and the angels of the vital fluid and in their cooperation all came out as it ought to be.

The dandy expressed amazement at this mystical text that he admitted he could not place. Shmuel explained that since he was a young student, he would have to wait weeks before the rebbe would see him. He said, "Since you are such an important man, you will be invited to see the rebbe soon after you arrive in town. Please ask the rebbe about the text and tell me what he says." The dandy agrees and does get to see the rebbe without a long wait. The rebbe slowly reads from the scrap of paper, closes his eyes and stares into inner places searching to see how the text makes sense. He opens his eyes and turns to the anxious dandy explaining the text with one word: *kreplach* (a Jewish version of ravioli).

"In the very primal beginning there was chaos – all was sundered and separate, grainy nuclei unconnected swirling." (That was flour.) "Then fiat, they were in one sphere." (Dough.) "The sphere unfolded into an orb." (The dough was rolled out flat.) "On the orb, lines appeared, forces cut the space into fields." (Of course – diamond-shaped pieces of dough are cut and meat put in.) "The fields became centered in a point and enfolded the point. Peace was made between fiery angels and the angels of the vital fluid" (as the pot was filled with water and put on the stove to boil, the kreplach were put in) "and in their cooperation all came out as it ought to be."

The rebbe laughed when he finally saw Shmuel. "What a dish you cooked up," he said.[12]

Creative play with biblical texts to glean new insights is a traditional method of Torah study. Playful spirit ingrained in those with Torah consciousness, makes it significant that my son Ari, publisher of the acclaimed website www.weeklyecho.com, taught me his creative interpretation of Bezalel's lineage. If we look at the name Betzalel, son of Uri (light) and grandson of Hur (freedom), in a conceptual order rather than its linguistic order, the names are reversed. Chronologically the grandfather comes before the son and grandson. In an artist's process of working, freedom gives birth to light. Exploring the parallelism between the artist Betzalel's creativity and God's, Ari read the beginning of the Bible in the original Hebrew. "The earth was astonishingly empty, with darkness on the face of the depths, and God's spirit hovered over the water's surface. God said, 'Let there be light' and there was light" (*Genesis 1:2, 3*). Betzalel means "In God's Shadow," telling us that the artist creates light in the shadow where God's light is weak, just as God creates in a dark void. Following his reversing the order of the names of Betzalel's father and grandfather, Ari reversed the order of the letters in the words "spirit hovered," *ruah m'rahefet*. It becomes *hur tapah ram*, "freedom greatly swelled (like dough rising)." "God's freedom powerfully rising up to create light" parallels Hur (freedom) giving life to Uri (light). Jewish material culture celebrates freedom from Egyptian bondage each Sabbath by eating *ḫalah*, puffed-up spiral loaves of bread, in contrast with *matzah*, flat unleavened bread eaten at Passover to symbolize bread of affliction.

Being created in the image and likeness of God is understood as an invitation to each human being to be God's partner in the creation of new worlds. The Bible follows the story of creation with the seventh day on which God rested from all creative work in order for humanity to continue the process on the eighth day. The biblical passage reads, "that God ceased from all the work which He had been creating *to do*" (*Genesis 2:3*). It would seem that the sentence would end after the word "creating." The word *la'asot*, "to do," however, is added to teach that we realize our being in the divine image by acting creatively, by continuing the divine pattern of creation, by bringing into being something new and original. According to the major 20th-century theologian, Rabbi Joseph B. Soloveitchik, the first and central biblical commandment is that each person is "obligated to engage in creation and the renewal of the cosmos." [13] The dream of creation is the central idea in biblical consciousness, "the idea of the importance of man as a partner of the Almighty in the act of creation, man as a creator of worlds."[14]

Human creativity finds its divine parallel in the scriptural passage, "God founded the earth in Wisdom. He established heavens in Understanding. With His Knowledge the depths opened and skies dripped dew" (*Proverbs: 3:19–20*). The words Wisdom, Understanding and Knowledge are only found together in the Bible in relation to the Creator of the universe and the artist Betzalel. They represent the beginning cognitive stages in the creative process. The next six stages in the creative process are affective stages representing emotions. They are revealed in *Chronicles 1:29*: "Yours, God, is the Compassion, the Strength, the Beauty, the Success, the Gracefulness, even Everything in heaven and earth." The final stage is the Kingdom of time and space in which intentions, thoughts, and emotions are realized in the everyday world of action. These stages in the creative process are called the ten *sephirot.*

The kabbalistic model used to describe creative process was created in the 16th century in the Galilee mountain town of Safed by Rabbi Isaac Luria, better known as the Arizal. It is called a "Tree of Life." It is a graphic model of 10 stages, *sephirot* (singular *sephirah*), represented as circles interconnected by 22 pathways corresponding to the letters of the Hebrew alphabet. The kabbalistic model is a metaphorical way of thinking rather than a body of knowledge to be seized. It offers a choreography for a dance of the mind to be apprehended by the part of the mind that appreciates poetry and hears its inner music.[15]

The empirical data illuminating the dynamic process described by this kabbalistic model is derived from two sources: an insider's personal narratives and an outsider's objective research. The narrative describes my experiences as an artist creating art. I attempt to follow the biblical form for conveying concepts and values through *sipur*, story telling. The Bible teaches through telling the narratives of the patriarchs' lives and the struggle of a people longing to be free and to bring holiness into everyday life. The research focuses on my analysis of in-depth interviews of 20 artists and scientists outstanding in their fields. I interviewed artists who have their work represented in the permanent collections of major museums. The scientists I interviewed are distinguished professors, members of the US National Academy of Sciences and Nobel laureates.

Eduardo Kac's collaboration with biologists is just one of many examples of the blurring of the lines between art and science in the postdigital era. I, therefore, juxtapose artists' statements with those of scientists. I asked each of them to describe aesthetic experiences that led to their creation of significant new scientific concepts and artistic statements. Through content analysis of the interview transcriptions, categories of experience are established that parallel the ten *sephirot* in the kabbalistic model that I use to describe the dynamics of creative process. I have found that the kabbalistic model possesses a greater degree of elegance than the ecological model that I had developed earlier by integrating psychological, biological, and mathematical models. This ecological model and the full interview transcriptions of the 20 artists and scientists can be found in my book, *Aesthetic Experience in Creative Process.*[16]

Ten *sephirot* of creative process

The process of creating a series of artworks, *Subway Angels*, illuminate a kabbalistic model composed of 10 stages (corresponding to 10 *sephirot*) and 22 pathways between them (corresponding to the 22 Hebrew letters). The description of this creative process emphasizes the flow from stage to stage, from preconscious intention to realization in the physical world. It provides a framework for understanding from a kabbalistic perspective how *Subway Angels* emerged as part my *Digitized Homage to Rembrandt* project. Following this narrative that spanned several years, I dwell in greater depth on each of the ten *sephirot* separately. Each *sephirah* (singular of *sephirot*) is illuminated by the creative experiences of prominent artists and scientists that they described to me in recorded interviews.

The creative process begins in the preconscious realm of faith, pleasure and intention. This first of the *sephirot* is the *sepherah* of Crown (*Keter*). Just as a crown floats above the head, so Crown is an undifferentiated longing to create that precedes the cognitive realm. It sets the stage for the first flash of insight. It is made up of three elements – faith that one can create, anticipation that the creative process is pleasurable, and intention to create. Without this self-confidence, hope for gratification, and will to create, the creative process has no beginning.

The inner experience of the second *sephirah* of Wisdom (*Hokhmah*) requires a selfless state. The nullification of the ego opens gateways to the subconscious. When active seeking ceases, when consciously preoccupied with unrelated activities, when we least expect it, the germ of the creative idea bursts into our consciousness. This sudden flash of insight is what the *kabbalah* calls wisdom. It is the transition from nothingness to being, from potential to the first moment of existence. In biblical words, "Wisdom shall be found in nothingness" (*Job 28:12*).

The process of creating *Subway Angels* began as I sat in a small Hasidic synagogue in Brooklyn following the reading of the weekly biblical portion from the handwritten Torah scroll. I listened to the ancient Hebrew words, translating them into English in my mind. The sentence about the prototypic artist Betzalel being filled with divine spirit, wisdom, understanding and knowledge, and talent for all types of craftsmanship is followed by "to make all manner of *MeLekHet MakHSheVeT*" (*Exodus 35:33*). Usually translated as "artistic work," it literally means "thoughtful craft."

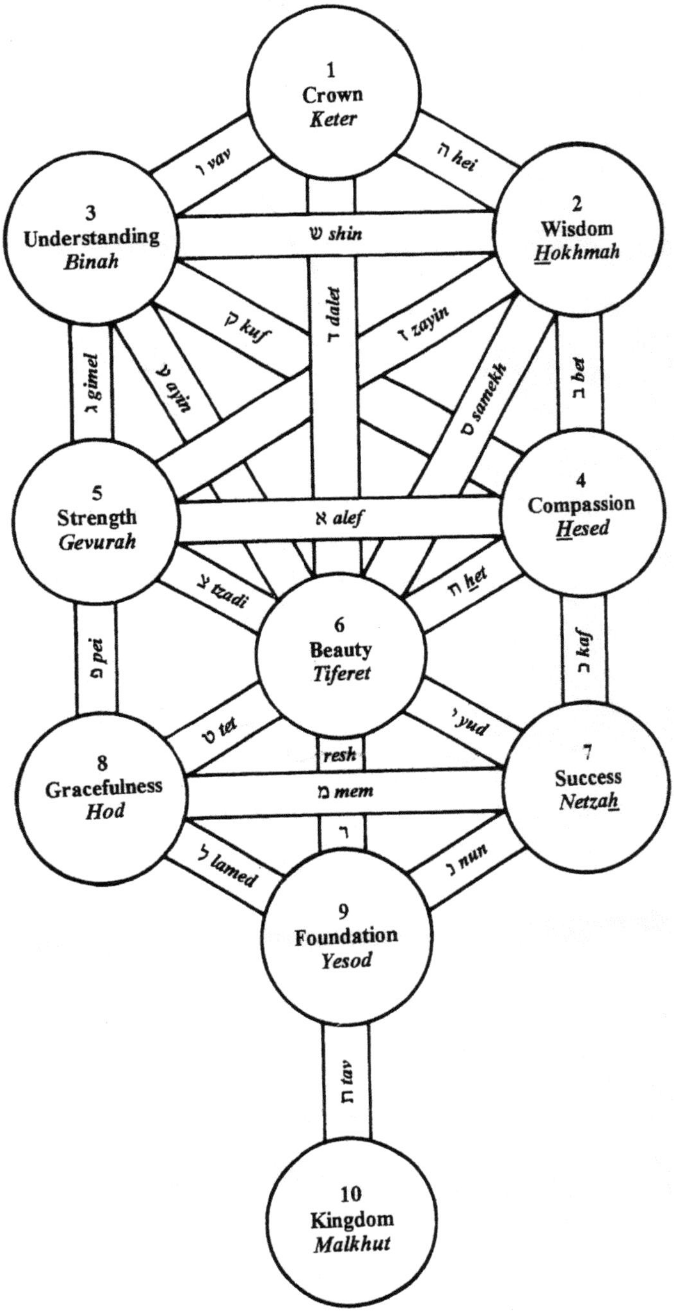

Figure 2: Kabbalistic model of creative process.

At that moment, I was living in the Crown *sephirah*. As an artist, I subconsciously intended to create artworks; I had faith in my ability to create artworks; and I felt that I would derive pleasure from the process of making art. However, it was the Sabbath and I was removed from my studio, from my classroom where I taught computer graphics, and from my office as head of the art department at Pratt Institute. Indeed, the definition of Sabbath rest is to refrain from making *MeLekHet MakHSheVeT*. The Sabbath day is biblically defined as the Non-Art day. It is the day in which all work on the Tabernacle was suspended. To this day, an observant Jew on the Sabbath avoids doing any of the 39 categories of thoughtful craft that went into the biblical artists' creation of the Tabernacle.

My absorption in the rhythm of the chanting of the Torah put me into a meditative state. I was passively listening, open to receiving. The stage was set for the Wisdom *sephirah*. In a flash of insight I realized that as a male artist, I needed to create computer angels. It suddenly dawned on me that the biblical term for "art," *MeLekHeT MakHSheVeT*, is feminine. Its masculine form is *MaLakH MakHSheV*, literally "computer angel." Art is a computer angel when biblical Hebrew meets modern Hebrew in a postdigital age.

Like the sperm that is received by the ovum in the womb, the unformed germ of an idea from the Wisdom *sephirah* enters into the *sephirah* of Understanding (*Binah*). This union of Wisdom and Understanding is Knowledge (*Daat*), as Adam knew Eve. As soon as the synagogue service came to an end, I explained to my wife that I needed to make computer angels. "You need to make what?" she responded incredulously. As I transformed my unformed insight into words to explain my thoughts to her, I entered into the Understanding *sephirah*. All manner of thoughts entered my mind on ways to create computer angels. The shapeless idea that ignited the process began to take form in the Understanding *sephirah*. Together, Wisdom, Understanding, and Knowledge form the cognitive realm of thoughts. Knowledge both unites Wisdom and Understanding and is the gateway to the next six *sephirot* that form the affective realm of emotions.

The fourth *sephirah* of Compassion (*Hesed*) is openness to all possibilities. I thought of the hundreds of artistic options open to me in creating computer angels and I love them all. Compassion is counterbalanced by the fifth *sephirah* of Strength (*Gevurah*), the strength to set limits, to make judgments, to choose between myriad options. It demands that I make hard choices about which paths to take and which options to abandon. What angel images do I digitize? What media do I use? Should I make paintings, lithographs, serigraphs, etchings, multimedia works, videos, or telecommunication events in which cyberangels fly around the planet via satellites? Even if I would love to do all of these things, I must choose one of them to begin with.

I recalled that a few weeks earlier, my son Ron had sent me an article on Rabbi Kook's views that the light in Rembrandt's paintings was the hidden light of the first day of Creation.[17] At the time, Ron was archivist at Beit Harav Kook in Jerusalem, the residence of the late kabbalist and chief rabbi of the Land of Israel, Abraham Isaac Kook. This article coupled with my connection to Holland through my wife's Dutch family made it clear that my first choice was to digitize Rembrandt's angels. I thought I should go to the print room at the Metropolitan Museum of Art where I could look at original Rembrandt drawings and etchings and select images of his angels. I knew that he had created a host of artworks of Jacob's dream. "A ladder was standing on the ground, and its top reached up toward heaven, and angels were going up and down on it" (*Genesis 28:12*). Since angels first go up before they go down, they must start their ascent from the lowest of places. In New York City, perhaps angels fly up from the subways. It seemed like a great idea to paint on subway posters and silk-screen print on them digitized Rembrandt angels and spiritual messages from underground.

As I felt deep satisfaction with my choice, I departed from the Strength *sephirah* to the next stage, the sixth *sephirah* of Beauty (*Tiferet*). This *sephirah* represents a beautiful balance resonating between the counter forces of Compassion and Strength. It is the feeling of harmony between all my possible options and the choices I had made. It is the aesthetic core of the creative process in which beauty, splendor, and truth are experienced as being intrinsically valued. Beauty is not a *sephirah* of static harmony, but rather a harmonious feeling shaped by the dynamic interplay of counter forces where Compassion does not overwhelm Strength and Strength does not overwhelm Compassion. I experienced the splendidly harmonious feeling that all is going beautifully.

The seventh *sephirah* of Success (*Netzah̲*) is the feeling of being victorious in the quest for significance. I felt that I had the power to overcome any obstacles that may stand in the way of realizing my artwork. The word *Netzah̲* is also related to "conduct" or "orchestrate" as in the word that begins many of the Psalms. I had the confidence that I could orchestrate all the aspects of creating a multimedia symphony of computer angels arising from the bowels of New York City.

The eighth *sephirah* of Gracefulness (*Hod*) is the glorious feeling that the final shaping of the idea is going so smoothly that it seems as effortless as the movements of a graceful dancer. The Success *sephirah* is an active self-confidence in contrast with the Gracefulness *sephirah* which is a passive confidence born of a trust in divine providence that "all will be good." It is the power to advance smoothly with the determination and perseverance born of deep inner commitment. The word *Hod* connotes both "acknowledgement" (*hoda'ah*) and "splendor" in the sense of an aura-like "reverberation" (*hed*) of light. The Gracefulness *sephirah* serves to endow the source of one's inspiration with an aura of splendor and majesty. It is the wonderful feeling that all is going as it should be.

The ninth *sephirah* of Foundation (*Yesod*) is the sensuous bonding of Success and Gracefulness in a union that leads to the birth of the fully formed idea. It funnels the integrated forces of intention, thought, and emotion of the previous eight *sephirot* into the world of physical action. In *Chronicles 1:29*, this *sephirah* is called All or Everything (*kol*). It channels everything that was playing out in my mind into the craft of making the artwork. It transforms my private mental world into a public environmental product that can communicate my ideas to others.

This tenth *sephirah* of Kingdom (*Malkhut*) is the noble realization of my concepts and feelings in the kingdom of time and space. It involves all the practical details that go into physically making an artwork. I began the actualization of my concepts by going to the company that places advertising posters in subway cars. They gave me fifty different placards on which I painted and silk-screened printed angels and spiritual messages based upon Hebrew word play.

Exhibiting my series of *Subway Angels* was a culminating activity that gave me the opportunity to stand back and look at what I had done. This activity is parallel to the divine act on the seventh day when God looked at the completed creation and saw that it was good. My sense of satisfaction, however, began to turn into a feeling of postpartum emptiness. I had given over my creations to the world and they were no longer mine to possess.

The tenth Kingdom *sephirah*, the realm of physical reality, was being transformed into the first Crown *sephirah*, the return to subconscious longing to create anew. The process had come full circle. The Kingdom and Crown *sephirot*, the end and the beginning, merge along an endlessly flowing Mobius strip. The ancient kabbalistic text, *The Book of Creation* (*Sefer Yetzirah 1:7*), teaches: "Their end is imbedded in their beginning and their beginning in their end like a flame in a burning coal." The linear progression through ten *sephirot* presented above is oversimplified to emphasize the flow from subconscious intention to actualization in the material realm. However, not only does the end transform itself into the beginning as in the Torah scroll, but there is movement in multiple directions between the *SePhiRot* and within them. The creative process is recursive and holographic. This complex multidirectional process can best be understood by hearing the narrative relating the story (*SiPuR*) of the artist's life experiences and thoughts. The narrative presents ways to see sources that shape the pool of images and ideas from which new artworks emerge. An exemplary *sipur* begins with my encounter with Rembrandt and Anne Frank in Amsterdam.

Digitized homage to Rembrandt

I walked across Amsterdam along the canals from the house on Joodenbreestraat (Jewish Broadway) where Rembrandt had lived to Westerkerk (West Church) where he was buried. There was no tombstone in the church courtyard marking his grave. No sign in or around the church indicated that it was the final resting place of the great master. On the sidewalk in front of the church, however, a bronze life-size statue of Anne Frank stood watch. She had been hiding in a room overlooking the church courtyard until the Nazis discovered her and carted her off to Bergen-Belsen to die. A postcard reproduction of a Rembrandt painting of an old Jewish man that she had tacked to the wall remained behind.

From Westerkerk, I took a tram back to my mother-in-law's apartment. I had traveled with my wife, Miriam, to Holland to be with her family during the *shiva*, the seven-day period of mourning for her father. He had suddenly died of a heart attack in Suriname, the former Dutch colony in South America where Miriam was born. It was the first time I had been outside of the United States. People who came to pay their respects told Miriam how lucky she was that her father had died a natural death, unlike her grandfather and her grandmother and her aunts and her uncles and her cousins. The Nazis murdered them all. Not one family member that stayed in Holland survived.

I was fortunate to have been born in Brooklyn in 1937, the year the aspiring artist Hitler launched the most virulent attack ever mounted against modern art by opening the exhibition *Entartete Kunst* (Degenerate Art) in Munich. This show was aimed at the defamation and derision of new directions in art that he called Jewish art even when Gentiles made it. His aesthetic solution for restoring the health of European folk was to rid the world of Jews and degenerate modern art. He almost succeeded. It is symbolic of the moral bankruptcy of Europe that the French and German Nazis herded the Jews of Paris into the courtyard of the Louvre. From that shrine to European culture, they shipped them to gas chambers and crematoria. Perhaps Duchamp deeply understood some of the darker messages of premodern European art when he drew a mustache on Mona Lisa and exhibited a urinal as a work of art.

My childhood memories of first seeing Rembrandt paintings at the Metropolitan Museum of Art were tied to horrific images of a Jew being murdered. The dread of being lost in the maze of rooms leading to the Rembrandts made my heart beat in fear. Each room was filled with frightening pictures of a young Jewish man suffering, a wreath of thorns crowning his bloodied head, gangrenous hands pierced with spikes, Roman soldiers nailing him to a cross for the crime of being called King of the Jews. Mixed with execution scenes were the pietas, paintings of an anguished Jewish mother holding her dead son. The varnished umbers cast a dark and dreadful glow. I sensed in horror that these pictures were visual lessons instructing people to torture and kill Jews. They screamed out, "Kill him! Kill Mel Alexenberg, Menahem ben Avraham ben Mordecai ben Elhanan, the Jew!" I raced through the museum like a gazelle fleeing a hungry lion until I fell into the welcoming arms of Tahitian women. Out of breath, I was comforted by their soft bronze breasts and the fragrance of the flower petals in the baskets they held out to me. The room with Gaugin's paintings was my hideout, my sanctuary. I felt the serenity of Gaugin's tropical colors wash over me, cleansing me, reviving me.

I relived my childhood terror four decades later as I walked through the Met to the printroom to select images of Rembrandt's angels to digitize. What I sensed as a child was true. From the Crusades to the Inquisition to the pogroms to the Holocaust, Europeans learned their lessons well. They drenched their continent with Jewish blood. The South Pacific light in Gaugin's paintings saved me from the European darkness. I recalled reading how Joseph Goebbels, Reich Minister of National Enlightenment and Propaganda, led Hitler through a warehouse of "degenerate art" to convince him to sell off the "garbage" rather than burn it. An auction was arranged in Switzerland. Among the 15 paintings bought by the Belgians was Paul Gaugin's *From Tahiti*. Gaugin's "deranged vision" was traded for Nazi guns.[18]

When I reached the printroom, I was seated at a large oak table. In a quiet ritual, one Rembrandt at a time was placed on a delicate easel in front of me as the tissue paper protecting the picture was slowly removed. As his etching *Abraham Entertaining the Angels* was uncovered, I saw that only two of the angels had wings. The figure facing Abraham had no wings. Perhaps Rembrandt wanted to show that although they looked like men to Abraham, they were really angels in disguise.

The Torah (*Genesis 18:1–8*) relates how three angels disguised as men appeared to the Abraham while he was sitting at the entrance to his tent in the heat of the day. When he looked up and saw them a short distance from him, he ran to greet them and invited them to eat with him. He rushed to his wife, Sarah, and asked her to bake cakes for their guests. Then Abraham ran to the cattle to choose a tender, choice calf. The *midrash*[19] questions why Abraham ran after the calf. The calf ran away from him into a cave. When inside, he discovered that he had entered the burial place of Adam and Eve. He saw intense light emanating from an opening at the end of the cave. He was drawn to the light. As he approached, he saw the Garden of Eden through the opening. This deeply spiritual person, the patriarch Abraham, found himself standing at the entrance to Paradise. About to cross over the threshold into the pristine garden, he remembered that his wife and three guests were waiting for lunch back at the tent. What should he do? Should he trade Paradise for a barbeque? The Torah tells us that he chose to return to the tent and join his wife in making a lunch for the three strangers. They sat together in the shade of a tree and enjoyed the barbeque. We learn from this legend that we ourselves create heaven or hell in our relationships with our spouses, children, friends, neighbors and strangers. Visions of Paradise far off at the end of a cave or in some heavenly realm above are mere mirages or fraudulent lies. Abraham knew that he and Sarah had the power to create heaven together in their tent.

The museum supplied photographs of etchings and drawings of winged people that Rembrandt had made to represent angels. I digitized them and began to manipulate them with computer graphics programs with the intention of using them in paintings and prints. My sudden interest in computer angels came at a time when I was involved in making a series of paintings of storefronts. I got into storefronts as a result of a discussion with Louise Nevelson on the ugliness of Brooklyn. After living for seven years in the bright light of the Negev mountains, finding myself in Brooklyn gave me aesthetic blues. The Brooklyn sky looked sidewalk gray. The sidewalks were dirty, the buildings drab. I missed the flowers that bloomed beside the Negev streams after the first winter rain: red anemones, poppies with paper-thin petals, black irises with sun-yellow cores, and clusters of bell-shaped flowers named *iyrit,* my daughter's name.

I met Louise Nevelson in her elegantly furnished home on Mott Street, where SoHo meets Chinatown and Little Italy. As head of the art department at Pratt Institute, I had come to invite her to speak at commencement. While complaining about my unsightly neighborhood, she pointed to a rocking chair across from where I was sitting. She told me about an art critic who had come to interview her for *ARTnews* and had the chutzpah to ask her why she owned such an ugly, kitsch rocking chair. Louise lectured me in her deep voice, "I told him that he should see the amazing shadows that the rocker casts each morning when the sun streams in. Mel, you need to be receptive of subtle bits of beauty, and they will jump out at you even on Brooklyn streets."

They did not quite jump out at me. They floated towards me in slow motion as in a dream. Maybe more like the after-image formed when you stare at a green, black, and orange American flag and see a red, white, and blue one when you look away. It happened early one Sunday morning while I was out on Avenue J buying fresh-baked bagels and the Sunday paper. For some reason, I turned around as I left the bagel shop. I stopped and stared at the storefront as if I had seen it for the first time. Neon Hebrew words danced above the food-filled windows.

I rushed home, ate breakfast, and returned to Avenue J with my camera to photograph the food stores. Next to the bagel shop was Isaac's kosher bakery topped with the words *chalav yisrael* (milk under rabbinical supervision all the way from cow to cake). In the three blocks between the train tracks and Coney Island Avenue, there were more bagel shops, kosher meat markets, kosher fish markets, kosher cheese stores, kosher take-out food places, kosher doughnut shops, and fruit and vegetable stands run by Jewish immigrants from Odessa. I photographed two kosher Chinese restaurants with oriental-sounding names: Glatt Chow and Shulchan Low (*shulchan* means table in Hebrew, *glatt* is a Yiddish word referring to "unblemished lungs," a sign of especially kosher meat). I finished my roll of film on three kosher pizza parlors each named for a different city in Israel: Netanya Pizza, Jerusalem Pizza, and Haifa Pizza.

After I finished Avenue J, I went kosher-store hopping throughout Flatbush, Boro Park, and Crown Heights. I photographed more than 100 storefronts. It would seem that Judaism was about food. Kosher food stores were far more conspicuous than synagogues tucked away in what appeared to be private homes. These stores, crowned with Hebrew neon, seemed to me to be strangely out of place. They looked as if they had been plucked up from a street in Israel and plopped down in America by a band of mischievous angels.

I enlarged some of the storefront photographs on a copy machine and then repeated enlarging the enlargements until they were three feet high. Since people do not usually stop and stare at storefronts but walk by them, I cut the images of a row of storefronts in strips, repeated the images to give the feeling of movement, and glued them onto a Masonite panel. I painted over the fragmented storefronts with layers of acrylic paint creating tension between the hand-made quality of the textured surface and images generated by a copy machine. I also made oversized Kodaliths (high-contrast negatives) of other storefronts and silk-screened printed alternating images of negatives and positives on canvas. Two rows of storefronts were printed stacked horizontally as if the viewer was hovering over the street and could see both sides of the street at the same time. I painted over the black printing ink with bright acrylic colors. Although the paintings looked finished, subconscious nagging told me that something was missing, but I had not the slightest idea what it was. I wrapped them and stored them away.

It was a few weeks later, while hearing the Torah reading in Rabbi Rutner's modest synagogue on the ground floor of his home around the corner from Avenue J, that the *ḫokhmah* flash of insight revealed to me that "computer angel" was the masculine form of the biblical term for "art." Having abandoned my storefront paintings, I began working with digitized images of Rembrandt's angels. I created the *Subway Angel* series and lithographs, serigraphs, etchings, and other paintings of angelic activities.

On *Simkhat Torah*, the holiday celebrating the ending and beginning of the annual Torah-reading cycle, Miriam and our youngest son, Moshe, spent the celebration in Crown Heights with the Hasidic community of the Lubavitcher Rebbe. After the evening prayers, Torah scrolls dressed in velvet mantels and capped with silver crowns where removed from the ark. A scroll in a royal blue mantle embroidered with golden lions was placed in my arms. Moshe petted the lions and latched on to the bottom of the mantle as we danced away together. The floor rumbled, as the dancing grew swifter and the singing louder. We were swept away into a line of dancers flowing from the feverish air of the synagogue out into the chill of a Brooklyn night. Meandering through the crowded street, we soon found ourselves back in the synagogue. Moshe and I kissed the Torah as I handed it to a young man who danced away with it in his arms.

As I sat down to rest, a man with an unruly beard and ruddy tan complexion greeted me in Hebrew. He had come from Afula in Israel to be with the Rebbe for the holiday. He asked me what I did. I told him I was an angel-maker, an artist who created spiritual messages. I explained that the term for art in the Torah, *MeLAekHeT MakhSheVeT*, is the feminine form of computer angel, *MaLAkH MakHSheV*. He added that the word *MAakHL* (food) has the same letters as *MaLAkH* (angel). The biblical words for "angel" and "food" are written with the same four Hebrew letters to tell us that angels are spiritual messages arising from everyday life.

I immediately knew what was missing in my paintings of food store facades – angels! My new Wisdom-*Hokhmah* insight linked two disparate realms in what Arthur Koestler calls "bisociation" in his book on the creative process, *The Act of Creation.*[20] He makes the distinction between routine skills of thinking on a single plane and the creative act culminating in a new relationship formed at the intersection of two different planes of thought.

The man from Afula continued to explain that the gematriah of the Hebrew words for "angel" and "artist" both equal 91, mystically uniting them. Artists can be vessels to receive angels (divine inspirations). Artists can also be angel-makers, creating artworks that release new angels (spiritual messages) into the world for others to receive. He pointed out that the Hebrew word for "spiritual" is essentially the same word as "material" spelled backwards. If we shift our perspective, we can transform our encounters with the material world into spiritual ones. Judaism's goal is to make all aspects of our lives holy and our everyday world a dwelling place for God. "Wine, women, and song" is an expression for crass materialism among English speakers. In Jewish life, *kiddush* (sanctification) is the blessing made on drinking wine, *kiddushin* (holy vows of matrimony) is the basis for intimacy with a woman, and *Song of Songs* is called *kodesh kodeshim* (holy of holies), traditionally regarded as the most sacred book in the Bible. Judaism strives to transform the grossest materialism into the most refined spirituality.

Returning home from our holiday in Crown Heights, I pulled my abandoned storefront paintings out of storage. I understood that bringing computer angels into these paintings would raise them to a new level of significance. They would express Hebrew linguistic connections between food and angel, between artist and angel, and between the material and spiritual realms. I glued printouts of digitized images of angels flying out of the storefronts in my paintings. I cut out large computer angels that had been lithographed on fine hand-made paper and pasted them onto the screen-printed canvases. They appeared to be hovering over the street between the two rows of stores. The paintings moved beyond being mere illustrations to becoming works of art that evoked fresh relationships between material and spiritual worlds. In addition, the playful juxtaposition of the mundane with the holy in Western art, kosher food stores in Brooklyn with Rembrandt drawings, created ambivalence between homage and contempt for art made by the hand of a master on the continent that created the culture of Auschwitz. In *Fixing the World: Jewish American Painters in the Twentieth Century*, with my painting of a cyberangel emerging from a storefront on Coney Island Avenue reproduced as the cover of the book, Ori Soltes writes:

> Alexenberg appropriates an iconic image from the Christian artistic tradition: Rembrandt's angel, who wrestles with Jacob. But he transforms and distorts it, digitalizing and dismembering it, transforming the normative Western tradition within which he works as he rebels against it.[21]

Cyberangels and AT&T

Working with Rembrandt's angels, reminded me of the small etching he had made as a book illustration showing angels going up and down the ladder in Jacob's dream. It was in the only book he had illustrated, *Piedra Gloriosa/Even Yakar* (*Glorious Stone* in Ladino and Hebrew), a kabbalistic book written by his neighbor and friend, Rabbi Menasseh ben Israel. I wanted to do something to honor Rembrandt. On October 4, it would be the 320th anniversary of his death. Jewish tradition honors people on the day they complete their lives rather than on their birthdays. It is like applauding after seeing a great play instead of when the curtain opens. It dawned on me that I could applaud Rembrandt best by having his winged angels wing their way around the world.

I phoned AT&T. I asked if I could use their telecommunications satellites to send a cyberangel on a circumglobal flight. "You have *what* to send around the globe?" was the usual response as I was transferred from office to office. Incredulity was turned to interest when I reached the director of the Infoquest Center, AT&T's technology museum on the ground floor of their postmodern building designed by Philip Johnson. I took a clangorous subway train across the Manhattan Bridge to present my proposal. The public relations people liked the idea and AT&T agreed to sponsor my memorial fax-art event.

I flew to Amsterdam to meet with Eva Orenstein van Slooten, curator of Museum het Rembrandthuis, the artist's home and studio. With trepidation, I proposed having a fax machine placed on Rembrandt's 350-year-old etching press to receive the angel that would fly there from New York. She thought it was a wonderful idea. It would make her museum, a quiet place, come alive as Rembrandt's angel rematerialized in the place he had originally created it.

On the morning of October 4, his angel ascended from the Chippendale top of the AT&T building in New York. It flew to Amsterdam to Jerusalem to Tokyo to Los Angeles, returning to the former New Amsterdam on the same afternoon. It took an hour in each city to receive 28 pages of angel fragments and fax them on to the next city.

After a five-hour flight around the planet, the deconstructed angel was reconstructed for the fifth time at its starting point. When it passed through Tokyo, it was already the morning of October 5. After the line printed out on the top of the fax "Tokyo National University of Arts and Music, 5 October 1989" was the line "Museum of Contemporary Art, Los Angeles, 4 October 1989." Cyberangels cannot only fly around the globe, they can fly into tomorrow and back into yesterday. They reshape our concepts of time and space in ways that correspond to the vision of kabbalists centuries ago.

The cyberangel was received at Rembrandt's house seconds after it left New York. It came as 28 sheets, each with an abstract fragment of the angel image. Ms van Slooten fed the sheets back into the fax machine on Rembrandt's etching press and dialed the fax number of the Israel Museum in Jerusalem. She then assembled all the fragments into a whole 4 x 6-foot angel.

Jerusalem was the appropriate next stop since it is an angel from a biblical scene. It was evening when the cyberangel arrived. Amalyah Zipkin, curator of European art at the Israel Museum, sent me a description of the angel coming and going. She wrote:

> There is something appropriate in the illogic of the event: here we were in Jerusalem, the Holy City of 4,000 years of turbulent history, huddled next to a fax machine in the mail room of the Israel Museum. It was a few days before Yom Kippur. Somewhere out there in technological space, a disembodied angel – computerized, digitized, enlarged, quartered, and faxed – was winging its way towards us from Amsterdam. This angel had been drawn in the 17th century by a Dutch artist with the instantly-recognizable mass-media name of Rembrandt van Rijn, and had undergone its electronic dematerialization 320 years after the artist's death as the hands of a New York artist and technology freak who had the audacity to make the connections: Rembrandt, the Bible, *gematria*, the electronic age, global communications, the art world, and the fax machine. Like magic, at the appointed hour the fax machine zapped to life and bits of angel began to materialize in Jerusalem. Photographs and the attendant PR requirements of contemporary life were seen to, and the pages were carefully fed back into the machine. We punched in the Tokyo phone number and the angel took technological flight once more.

It was almost dawn on October 5 when the angel arrived in Tokyo in the Land of the Rising Sun where fax machines are made. Ikuro Choh of the Tokyo National University of Arts and Music received the angel and revealed its full image by assembling the 28 sheets on the ground among the ancient pillars in Ueno Park. He then disassembled them and attached all the sheets end-to-end in a long ribbon ascending the stairs and entering into a centuries-old religious shrine built in traditional pagoda style. The old Tokyo site was selected to carry a spiritual message of electronic age homage to tradition. Ikuro Choh laments, "not only in Tokyo but everywhere in Japan, the traditional and the old are being destroyed at a ferocious speed, making the culture of paper, wood and bamboo evaporate like a mist, allowing the ugly demons of concrete to appear in its wake." With the sun rising over Japan to begin a new day, the fax-art angel rose over the Pacific Ocean to fly into yesterday. It arrived in the City of the Angels at 2:40 p.m. on October 4. The angel came together once again at the Museum of Contemporary Art in Los Angeles on the day before it had visited Tokyo.

The cyberangel returned to New York five hours after it had left. It had entered tomorrow before flying forward into yesterday. Camera crews for all the major television networks welcomed the cyberangel's return from its circumglobal flight. It was broadcast on the national news from New York that evening. After having flown around the world, the cyberangel simultaneously visited millions of homes across North America. Associated Press covered the fax-art event, too, sending the angel image and story over its wire services. Sixty newspapers carried the AP story, each with a different headline. It even made the front page in Billings (Montana), Marion (Ohio), and Selby (North Carolina).[22] AT&T made it the feature of their annual report.[23] They distributed three million copies showing me, a gray-bearded Jewish artist sporting a Hasidic black hat, welcoming the cyberangel on it return from its high tech flight around our planet.

Lucy Lippard's words in *Mixed Blessings: New Art in a Multicultural America* call for repetition here to best summarize the postmodern concept behind the computer angel story: "I am interested in cultural dissimilarities and the light they shed on the fundamental human similarities," as well as "art that combines a pride in roots with an explorer's view of the world as it is shared by others."[24]

Creativity in art and science

The kabbalistic model of creative process is universally applicable, it extends beyond art to all human creativity. Biblical consciousness recognizes two creation stories to emulate: God's creation of natural systems and the human creation of the Tabernacle. Science derives from the former and art from the latter. There are also two biblical creative endeavors that are disastrous: the Tower of Babel and the Golden Calf. They are warnings that human creativity can be used for malevolent purposes as well as for good.

As we learned earlier, the first time that the Hebrew root *KBL* of the word *KaBaLah* appears in the Bible is in the word for parallel, *maKBiL*. Divine and human creativity are parallel to one another. Science and art are two parallel human endeavors that share the aesthetic dimension, the Beauty *sephirah*, at the core of their creative processes. Although the link between aesthetics and art is obvious, the link between aesthetics and science is not often acknowledged.

Abraham Maslow studied the creative process in science and found that there is a shift in attitudes in science from attempts to purge itself of transcendent experiences to acknowledging that enjoyment of acquiring knowledge is the richest source of aesthetic experience and feelings of awe and mystery. He writes in *The Psychology of Science*: "Such experiences of joy are necessary not only because they bring people into science and keep them there, but also because these esthetic joys may also be cognitive signs, like signal rockets that go off to tell us we have found something important."[25]

Albert Einstein not only connects art and science in their shared aesthetic emotion of standing in awe before the beauties of this world, but also links this aesthetic experience to Jewish consciousness as expressed in the Psalms. He writes: "The fairest thing we can experience is the mysterious. It is the fundamental emotion which stands at the cradle of true art and true science. He who knows it not and can no longer feel amazement is as good as dead, a snuffed-out candle."[26] He sees the central aim of Judaism as the "sanctification of life in a supra-personal sense." He continues:

> But the Jewish tradition also contains something else, something which finds splendid expression in many of the Psalms – namely, a sort of intoxicated joy and amazement at the beauty and grandeur of this world, of which man can just form a faint notion. It is the feeling from which true scientific research draws its spiritual sustenance, but which also seems to find expression in the song of birds.[27]

I believe that what we call art and science is integrally one in the human psyche. They only need to be merged in postmodernism because they were artificially torn apart. As a child set free among the salamanders and swallows in the Catskill Mountains every summer, I was engaged in what could be called scientific and artistic activities without having a clue that they were considered different areas of human endeavor. I studied the salamanders' habitat and behavior patterns: what they ate, how they interacted with their environment, each other, and animals that were their enemies and friends. I made drawings documenting my discoveries. I built fantasy salamander houses and villages. I designed terrariums and aquariums for the different species of salamanders and newts. I invented imaginary salamanders of vibrant pinks with green dots and made paintings of them. My childhood summers playing in the woods set the directions for my adult explorations, first as a biologist and science educator, and later as an artist and professor of art. The general systems theory that informed my scientific work on the ecology of terrestrial isopods continues to inform my quest as an artist. I am inspired by the merging of art and science in the genetic art of Kac and the ecological art of the Harrisons.

The scientific impulse in biblical consciousness finds its origin in the Crown *sephirah*. It begins as Moses was tending sheep in a mountainous region of the Sinai desert. In *The Living Torah*, Rabbi Aryeh Kaplan translates passages in *Exodus* from the original Hebrew into contemporary English:

> God's angel appeared to [Moses] in the heart of a fire, in the middle of a thorn-bush. And he looked, [Moses] realized that the bush was on fire, but was not being consumed. Moses said [to himself], "I must go over there and investigate this wonderful phenomenon. Why doesn't the bush burn?" When God saw that [Moses] was going to investigate, He called to him from the middle of the bush. "Moses, Moses!" He said. "Yes," replied [Moses]. "Do not come and closer," said [God]. "Take your shoes off your feet. The place upon which you are standing is holy ground." [...] "Right now the cry of the Israelites is coming to Me. I also see the pressure to which Egypt is subjecting them. Now go. I am sending you to Pharaoh. Bring My people, the Israelites, out of Egypt." [...] Moses said to God, "So I will go to the Israelites and say, 'Your fathers' God sent me to you.' They will immediately ask me what His name is. What shall I say to them?" "I Will Be Who I Will Be," replied God to Moses. [God then] explained, "This is what you must say to the Israelites: '*I Will Be* sent me to you.'"[28]

Moses sees a bush on fire, probably a common sight in his everyday world. When he realized that the fire was not consuming the bush, his scientific curiosity urged him to investigate this anomalous phenomenon. He is told to remove his shoes and stand in a different place to shift his viewpoint, to see things from a new perspective. Egypt is pressuring the Israelites to accept their enslavement. (Egypt in Hebrew is *Mitzrayim*, "narrow straits.") Moses is commanded to save his people from being pressured into narrowness of thinking, from being brainwashed into accepting a life without freedom. Moses is told to tell the Hebrew slaves that His name is *Ehyeh asher Ehyeh*. He is to tell them, "*Ehyeh* sent me too them." The Hebrew word *Ehyeh* begins with the letter *alef* which changes the verb "to be" to the future tense, to "I Will Be." God's name is a verb, not a noun, that actively points to a future open to all possibilities. Only if we are released from being enslaved in a world viewed from a narrow perspective, can we be free to create new and better worlds. Lastly, we learn that we can find holiness in a humble thorn-bush, in a lowly desert scrub. We can learn to hear God's voice in everyday encounters if we look closely from fresh perspectives.

Each of the ten *sephirot* has a divine name associated with it. *Eyheh* is associated with the Crown *sephirah*, *Shekhinah* with the Kingdom *sephirah*, and *YHVH* with the Beauty *sephirah*. *Eyheh*, "I Will Be," opens Crown to a future rich in possibilities. *Shekhinah* is the divine indwelling presence or immanence in Kingdom, the material world of space and time. She permeates all existence, even a wind-worn desert shrub. *YHVH* is "Is-Was-Will Be," the interplay between past, present, and future revealed as beautifully dynamic harmony. Crown is an empty vessel, the subconscious source of the insight that bursts into consciousness in Wisdom, the next stage of the process. The Hebrew word for Crown, *Keter*, can mean "circle" like a crown that encircles the king's head. It represents the beginning of the creative process emerging full circle from the end of an earlier creative process that had culminated in Kingdom.

Often fresh insights emerge during casual non-directed movement through everyday life removed from the worlds of work in art and science. Undifferentiated wandering through

Kingdom coupled with subconscious longing to create anew in Crown can trigger the transformation of a mundane experience into a brilliant new insight. Moses was wandering the desert hills with his sheep in a relaxed state of not looking for anything at all, when he encountered the anomaly of a burning bush that was not consumed. Donald Glaser and Will Barnet were able to solve problems in physics and painting through insights derived from casual encounters with ordinary events in their lives that they suddenly saw from new perspectives.

Donald Glaser's creative process that led to his being awarded the Nobel Prize in Physics began in Crown, the *sephirah* of faith, pleasure and intention. The starting point was his faith in himself as a scientist, his expectations that his scientific work would bring him pleasure, and his intention to solve key problems in physics. In his Nobel Prize acceptance speech, he said, "My own field of research, high energy nuclear physics, is especially remote from the experience of ordinary daily life since it deals with experiments on objects much too small to see or perceive directly." [29] However, his greatest insight came when he removed himself from the remote realm of high energy nuclear physics, focusing his attention instead on something as prosaic as observing what happens when he opens a soda bottle. When he eased up on the pressure in the bottle by removing the cap, he saw a rush of bubbles rising in the liquid. His insight came from his encounter with the ordinary daily life that he said was far removed from his work as a scientist. He thought how in high school chemistry we learn that carbon dioxide is an invisible gas. But it is not true. We can clearly see carbon dioxide as bubbles in soda. Now we are told that high energy subatomic particles are impossible to see. Like invisible carbon dioxide made visible in soda water, perhaps we could see the subatomic particles as bubbles if we smash an atom in a soda bottle. He asked "Why not?"

Glaser encountered the soda bottle in mundane Kingdom while his mind was empty of conscious scientific striving in the future oriented Crown of open-ended potential. When it dawned on him that it might be possible to see bubble streams made by subatomic particles in a liquid, he had entered the stage of Wisdom. Glaser could not have made the mindleap from soda bubbles to high energy nuclear physics without a prepared mind. Most people know that the bubbles in soda water are carbon dioxide, but they lack the reservoir of subconscious knowledge to make connections between disparate worlds. No intuitive flash of intellectual illumination will happen in Wisdom without serious preparation. Glaser considered himself a third-generation Nobel laureate. His major professor in graduate school was Nobel laureate Carl Anderson whose teacher was Nobel laureate Robert Millikan. He had earned his doctorate at the California Institute of Technology and was a professor of physics at the University of Michigan when the insight that led to his development of the bubble chamber occurred.

Following his question "Why not?" with initial ideas about how to make a bubble chamber for research in high energy nuclear physics, he entered into the Understanding *sephirah*, the expansion and elucidation of the Wisdom insight. Professor Siegbaum of the Swedish Academy of Sciences talked about this Understanding stage in the creative process in his speech on presenting the Nobel Prize to Professor Glaser:

> Glaser's next step in his line of reasoning was to use, instead of a soda-water bottle, a liquid which is heated to a temperature near its boiling point and which is enclosed in a chamber. If the pressure is rapidly eased from above the liquid, for example by use of a mobile piston, the liquid shall have a tendency to boil. With a certain amount of care, one is able to maintain the liquid in this superheated, unstable state without boiling. The slightest disturbance in the liquid, however, give rise to an instantaneous boiling exactly as did the opening of a soda water bottle. Glaser's idea was that an atomic particle passing through the liquid would be able to provoke boiling by means of the ions which the atomic particle produces along its path and which act as bubble development centers. If a flash picture were subsequently taken of the superheated liquid immediately after an atomic particle passes through, one ought to be able to observe the particle's path which is followed by the small bubbles which would have been produced.[30]

After years of his studying the physics of bubble formation both from theoretical and experimental points of view as well as technological advances, he was able to build a bubble chamber containing liquid hydrogen condensed by a large cooling device providing temperatures near absolute zero. The liquid chamber was surrounded by a powerful electromagnet that was capable of bending the paths of particles traveling close to the speed of light so that their curved bubbletracks could be photographed. He carried out his experiments on elementary particles at the Cosmotron of the Brookhaven National Laboratory in New York and the Bevatron at the Lawrence Radiation Laboratory in California.

Will Barnet is a painter whose works are in the collections of major American museums. He taught at Cooper Union and the Art Students League of New York with breaks as visiting professor at Yale, Cornell and Penn State. I studied painting with him at the League for two years. In my interview with him at Cooper Union, he describes the *sephirot* without any knowledge or reference to them.

> There are three stages that I sort of go through: First, I have a gnawing feeling that I want to do something, but I don't know what it is at that moment. And then the second one is where I became aware of what I want to do, but I have not yet found the right way of doing it. And then the third is when I begin to find real relationships and I begin to function to put things together. In other words, the concept comes together. But these are all mixed up with ideas of seeing something, feeling something, and then facing the white canvas. Doing all these things together. Synthesizing.[31]

When I asked him to describe the process of his involvement in making a painting, he chose to talk about a five-year process that culminated in his painting *Soft-Boiled Eggs.* After having spent years making drawings and small paintings documenting his children's activities around the house, he decided to make a major painting of his son's birthday party. "I had no clue at that moment as to just how I would handle it or how I would relate to it. I was still in what I call a fragmentary state." He made many drawings developing alternative ideas, most of which were unsuccessful. Finally the ideas began to clarify themselves for him. "To describe this picture there are three boys clustered around a woman, the mother. She's behind the table. One son's behind the table. One's underneath the table and one's on the left side of the table, standing on a ball in the final picture." Barnet described the many difficulties getting the figures to work with each other and in relationship to the table and the background. He felt emotionally elated when he finally resolved the relationships between the figures and their environment and experienced the wonderful feeling that the whole painting was coming together. To finish the painting, he only had to paint the birthday cake on the table. But the cake did not fit on the table. A single big circle placed in the center of the picture would ruin the delicate balance that he had created in the painting after months of hard work. He became very unhappy about it and put the painting away for a few weeks.

> Then, I came early one morning for breakfast. My wife had put a lot of soft-boiled eggs on the table, sort of arbitrarily. When I saw those eggs, I said to myself, "My God, that's the solution!" Because the eggs would not interfere with the horizontal movement of the table, they would become part of that whole plane. Then, the picture began to resolve itself. I still had the young child celebrating the birthday cake. He was celebrating soft-boiled eggs or celebrating anything – sort of universal. He was celebrating, period. It was no longer illustrative. Then, the painting became a painting and the story wasn't as important as the painting. But the story was still there."[32]

Barnet experienced a sudden revelation of the perfect solution by seeing an ordinary event in a refreshingly new way. He was in the Crown state of what Anton Ehrenzweig calls "undifferentiated conation" in *The Hidden Order of Art*.[33] Undifferentiated conation is the process of scanning an astronomical number of possibilities in a creative search. It involves advancing along a broad front in which contradicting options remain open while one is receptive to all possible combinations. In both the cases of Barnet and Glaser, their undifferentiated striving resulted in their locating their solutions in the Kingdom *sephirah*, in the realm of soda bottles and soft-boiled eggs. They were simultaneously functioning in the subconscious state of Crown and in the conscious state of Kingdom. The faith, delight, and intention of Crown were cast in an outward gaze at everyday life. This outward gaze triggered the Wisdom flash of insight. However, an inward gaze that does not depart from the subconscious Crown state can also trigger insight that bursts out from within. The broad scanning occurs within the mental realm without being directed outward into the environmental Kingdom. Past images and ideas are drawn up from memory to shape future creations in the artistic work of painter Henry Pearson and the scientific work of geophysicist Maurice Ewing.

Henry Pearson is one of the leading figures in Op Art, art that plays with mechanisms of visual perception. Op Art was defined by *The Responsive Eye* exhibition at the Museum of Modern Art in which Pearson's paintings were shown. I interviewed him in his storefront studio on the upper east side of Manhattan Island.

To support himself after earning his MFA at Yale, he got a job listening for flaws in phonograph records. Listening to the same records over and over again for three years left his hands free to make numerous drawings. His drawings developed into his abstract paintings exploring relationships between hard-edged rectangular masses. When his doodling led to curving the corners of his rectangles, he unexpectedly remembered the topographic maps of Japan that he had drawn as his job in the army many years before.

> For no reason at all, it suddenly occurred to me, because you always try things, to wonder how it would look if I were to curve the corners. I had never done that. So I tried curving the corners. In connection with the curving of the corners of the central rectangle, I curved the corner of the outer rectangle. Then, I remembered the topography and I said, "I wonder how it would be if I were to add another curved line after another curved line. I enjoyed the idea; so I made a few exploratory drawings with crayon at that time just to see what would happen. And then I went to ink and tried inventing landscapes of my own where the mountains were, where the valleys were. You could see how the lines would wonder in and out. Then, I felt that these were too realistic somehow. I wanted to make them more abstract [...] I was wondering how these drawings would look if I were to make them really large. But after I did the first one, I thought to myself, "That's pretty interesting. Maybe I should do some more of it."[34]

Pearson's relaxed state of listening to the same music repeatedly placed him in Crown, a state that lacks content, a mind-freeing state of meditation empty of conscious cognition, emotion and action. It set the stage for his being open to receiving a fresh connection between topographic maps and abstract painting as a spark of insight in Wisdom. As he began to think of developing his topographic doodles into paintings, he entered into the Understanding *sephirah.*

Maurice Ewing was America's foremost geophysicist. He founded the Lamont Geological Observatory of Columbia University and was its director until his retirement. He was the first recipient of the Vetlesen Prize that honors leaders in the earth sciences the way scientists in other fields are honored with Nobel Prizes. I interviewed him in his large office at the observatory moving through a labyrinth of paper pillars composed of stacks of charts, heaps of data records, and piles of scientific papers. When not pulling graphs for me to see from these pillars and checking ticker-tape communication from his research ships, he sat at a desk cluttered with mementos of his far-flung journeys while his aging dog sat at his feet.

Ewing had been studying relationships between ocean waves and the waves in the ground under the ocean by making explosions in the water and measuring the results with seismographs. He dreamed about exploring these relationships at the global level rather than in the local level. Ewing had no idea how to do it since there was no way he could make explosions great enough to shake the whole planet. The solution popped in his consciousness from his childhood memories. He recalled a story he had read in his fourth-grade reader about the great eruption of the volcano Krakatoa and suddenly realized that such an enormous explosion had actually occurred naturally. The eruption of a volcanic island between Sumatra and Java in 1883 made sea waves 50 feet high and killed nearly 40,000 Javanese. If he could have been there then and known in advance that such a massive explosion would occur, he thought, he would have placed tide gauges, seismographs, and barometers strategically around our planet. He could then have measured the waves in the ocean, waves in the ground under the water, and waves in the air above and established mathematical relationships between them on a global scale.

On a shelf in his laboratory, he had an old book, *The Eruption of Krakatoa and Subsequent Phenomena*, published in 1888 by the Royal Society of Great Britain. When he looked in this book, he realized that all the information he needed was there. It was fortuitous that the eruption occurred when Venus was eclipsed by the moon. Scientists throughout the world had placed tide gauges to measure the effect of the transit of Venus on the oceans. They knew that the alignment of the moon and Venus would cause extra-high tides. It was "beautiful and simple," Ewing said, to integrate information recorded by different people 70 years earlier into a unified theory of "resonant coupling." It was possible in one night to move into the study of surface waves, to understand it fully, to write the classic paper on it, to type it up, to put it in the mail. All in one night. "Well my friend, that is living. I don't know any thrill that anybody can have that will compare with that. Do you?"

The sudden linking of his elementary school reader's story on Krakatoa's eruption to his scientific research in geophysics was the first stage of cognition, the Wisdom *sephirah*. As he communicated his novel insight to his collaborator Frank Press, reached for the old book on his shelf, opened it, and saw the data he needed was there for the taking, he was functioning in the second cognitive stage, the Understanding *sephirah*. As the thrill and excitement of the process overwhelmed him, his consciousness flowed from the cognitive realm into the affective realm. The core of the affective realm is the aesthetic experience, the Beauty *sephirah*. Pathways from Compassion and from Strength connect to the central Beauty *sephirah* from above, as pathways from Success and from Gracefulness connect to Beauty from below. Beauty is reached as infatuation with multiple options is delimited so that success can be achieved gracefully. It centers human consciousness like a potter centering a lump of clay in throwing a pot. It is the inside of the pot around which emotions spin in exquisite balance. This balance, however, is not the static harmony valued by Hellenism, but rather a dynamic balance in which the pathways flowing into and out of Beauty are active processes engaged in aesthetic interplay.

Paul Weiss, head of the Laboratory for Developmental Biology at Rockefeller University, described Beauty as a dynamic balancing between the polarities of freedom and order. "The aesthetic experience is to me the attempt of man to depolarize extremes and reconcile them; to learn the dynamics of moving from the whole to the part and from the part back to the whole."[35] He explained that aesthetic experience fits between seeing freedom in the small with order in the gross. We can see that a maple leaf is clearly not an oak leaf. No two maple leaves, however, are congruent and share the same venation pattern. Each leaf displays order in the gross with freedom of excursion in the small. It is a unique, one-of-a-kind expression of a general pattern of a genetic type called a maple leaf. Aesthetic experience can depolarize the background of our biological past, which is polarizing opposites. To survive, an animal must polarize its view to differentiate between friend and foe and between food and poison. Humanity, however, can rise above the natural urge to classify experiences according to extreme properties. Aesthetic experience is the human way to reconcile antitheses.

The sculptor Ibram Lassaw described this dynamic balance in art by describing his experience making *Milky Way*, a sculpture that has been reproduced in many books on modern art. At the time, his work was an exploration of geometric relationships, although at an earlier time he had experimented with biomorphic forms. He made an armature, a skeletal framework of wire, and began covering it with a commercial plaster-like product. As he worked, the direction of his piece began to deviate from the geometric formality that he initially intended.

> It seemed to me as if something new was coming up in my work. It seemed like a real syntheses occurred between the biomorphic and the geometric kinds of structures, as though the geometric put on flesh or the biomorphic became crystallized. The two polar opposites in form seemed to fuse together in that piece. I was so happy and overcome by this feeling of having achieved something new, that something exciting was happening there, that I began dancing around my studio in sheer joy.[36]

Much of the inspiration for his artwork is derived from his ongoing encounter with scientific explorations of natural systems that reveal morphologies as well as direct observation of structures in nature. He had enjoyed trying to understand the work of scientists by subscribing to *Scientific American* for twenty years.

> I read more books on science than I do on art. I like looking at nature, seeing the shapes of trees and deriving enjoyment and understanding from it. I'm trying to understand the structure of the universe – the spiral galaxies, the galactic clusters, all the different things going on, the evolution of the universe, young stars making heavier elements of hydrogen, and all that kind of thing. It's the aesthetic comprehending of reality. The beauty and wonder of the living universe.[37]

Richard Lippold is best know for his wire sculpture, *The Sun*, in the collection of the Metropolitan Museum of Art and his monumental *Orpheus and Apollo* hovering over the entire entry lobby of the Lincoln Center's Philharmonic Hall. He dismisses things that are static and emphasizes how the constantly changing universe affects our lives and the artistic process. "If I sit down to make a sketch one day, surely it will be different from one I make another day. I feel different. We never feel the same from moment to moment; we are changing every moment. We are reshaping ourselves." He wrote in an article for the book *Structure in Art and Science* that "chance is an accident of order and order is an accident of chance." He described to me his making, *New Moonlight*, his first wire sculpture that became the ancestor of *The Sun* at the Met:

> The process involves the materials, the concepts, the feelings, and the physical making of the thing. Then anything can happen. It's charged with possibilities, with excitement, and with involvement. It's a little like making love. All the excitement is going on while it's going on. There is something comparable to an orgasm. It doesn't have quite this physical manifestation. But there is a sudden moment when I know it has come. It didn't mean to say it that way; it's the same word! Actually, this is what it is. You know that then everything has come together [...] The first time I made a work in which this happened, it happened quite mysteriously because I wasn't conscious of it. When the finished work arrived, appeared, the material and the technique came together in what struck me as an exquisite balance. The feeling didn't dominate; the concept didn't overwhelm; the form wasn't a technical exercise. Everything contributed to the meaning. It was a total balance which transcended all its parts, that together seemed a spiritual totality – an ecstatic event. It impressed me so that I couldn't sleep for a couple of days. I thought, "What have I wrought? I didn't determine all this. It happened through me; I was the medium for it. I was surprised. I was astonished, surprised and delighted.[38]

The biologist H. Bentley Glass compared aesthetic experience in science to the joy of discovery in geographical exploration and in musical composition. During his distinguished career in genetics, Professor Glass had served as president of the American Association for the Advancement of Science and the American Institute of Biological Sciences. He explained to me in his office at Stony Brook University on Long Island, the aesthetic value of students' involvement in real scientific experimentation.

> Nature is a hard master who reveals her secrets only to those who are patient and careful, persistent to the point of obstinacy and willing to spend time and energy without reckoning the cost. But if the student succeeds he will experience one of life's prize moments, a mixture of the joy of creation, like that of a composer who has just put the last note upon his score together with the exuberance of an explorer who from a jungle peak looks out over a vast ocean never seen before sparkling in the sun below and beyond.[39]

The importance of aesthetic joy in the scientific enterprise is echoed by Henri Poincare and by C. P. Snow, in their attempts to bridge the rift between the sciences and the humanities that has been artificially created by Western civilization. Poincare writes in *Science and Methods*: "The scientist does not study nature because it is useful to do so. He studies it because he takes pleasure in it, and he takes pleasure in it because it is beautiful. If nature were not beautiful it would not be worth knowing and life would not be worth living."[40] C. P. Snow proposes the existence of "Two Cultures" that are independent and often antagonistic. He demonstrates how the two cultures, the sciences and the arts, meet in their shared experience of aesthetic joy.

> Anyone who has ever worked in any science knows how much esthetic joy he has obtained. That is, in the actual activity of science, in the process of making a discovery, however humble it is, one can't help feeling an awareness of beauty. The subjective experience, the esthetic satisfaction, seems exactly the same as the satisfaction one gets from writing a poem or a novel, or composing a piece of music. I don't think anyone has succeeded in distinguishing between them. The literature of scientific discovery is full of this esthetic joy.[41]

The vital importance of the aesthetic experience in the scientific enterprise derives from the perception of beauty in nature and the feeling of joy at making a new discovery. These two aspects of aesthetic joy are viewed as a way to fulfill the biblical injunction, "Love God your Lord with all your heart, with all your soul, and with all your might" (*Deuteronomy* 6:5). According to Maimonides, one of the greatest Jewish thinkers of all time, the love demanded in the Bible can be achieved by observing and studying the Creation that God has gifted to humanity. The more the secrets of the natural world are revealed thought scientific research the more wondrous the Creation appears to us. Marveling at the beauties of nature reveals the divine plan and invites us to fall in love with the Master Planner. *Elohim*, God as the creator of the nature, and *hatevah,* nature, share the same numerical value of 86. Maimonides knew these aesthetic joys from his own scientific observations of nature. He was both rabbi and scientist who lived in 12th-century Spain and North Africa and was the personal physician to the Sultan. In his book, *Hilkhot Yesodai ha-Torah*, he extols science as a way to fulfill the *mitzvah* of *ahavat Hashem*, the commandment to love God: "When a person contemplates His great and wondrous works and creatures and from them obtains a glimpse of His wisdom which is incomparable and infinite, he will straightaway love, praise, and glorify Him, and exceedingly yearn and desire to know His great name."

In "Towards *Ahavat Hashem*: Ar t and the Religious Experience,"[42] Rabbi Chaim Brovender extends Maimonides' concept to the artistic enterprise. Beauty in nature can inspire us to express our relationship to God's creation through our own artistic creativity. The artist can gain entry into the special experience of God's love for humanity by reciprocating, by creating art that is a tribute to the Creator. God's gift of natural systems to humanity parallels His gift of spiritual media systems revealed through the Torah. "God has given us the Torah as an act of love, and we engage that love through the study of it." We learn from the major kabbalistic work, *Zohar*, that "God gazed into the Torah and created the world." Through engaging the blueprint and the end product, Torah and Nature, we can arrive at love of God. As an educator, Rabbi Brovender proposes that for those students who do not find the study of the Torah an entry-level experience into God's love, another pathway exists through art. "It might provide the necessary spiritual moment in an essentially nonspiritual existence that would enable them to begin to consider and later to experience God's love." Art can lead to experiencing the love of God, *ahavat Hashem*, by revealing the beauty in nature through the prism of human creativity.

Pathways to beauty

To best follow my discussion of four of the pathways to beauty, keep flipping between the text here and Figure 2, the diagram of the kabbalistic model on page 155. In the kabbalistic model, we have explored the significance of the *sephirot*, the ten stages in the creative process from intention to thought to feeling to action. The 22 pathways between the *sephirot* are the 22 letters of the Hebrew alphabet. As we discussed at the beginning of this chapter, the first divine creation is *et*, the first Hebrew letter *alef* connected to the last letter *tav* with the entire alphabet between them. To more fully understand beauty in Jewish consciousness it is instructive to look at the four Hebrew letters that form the four pathways linking Beauty with Compassion, Strength, Success and Gracefulness, the four *sephirot* that surround it in the world of emotions and feelings, known as the world of *Yetzirah*, the world of Creation. Three consecutive letters in the alphabet, the eighth, ninth, and tenth letters, *ẖet*, *tet*, and *yud*, form the pathways from Compassion, Gracefulness, and Success to Beauty. The pathway from Strength to Beauty is formed by the letter *tzadi*, the 18th letter with a numerical value of 90. We learn about each letter and the significance of its pathway through its number, form and name.

The cell, the basic element in living systems, provides a meaningful analogy for the *sephirot* in relation to the pathways leading in and out of them. The Beauty *sephirah*, exhibits what the biologist Waddington calls "homeorhesis" in his paper, "The Character of Biological Form."[43] He explains that organic form, whether the structure of a population, individual organism, or cell, "is produced by the interaction of numerous forces which are balanced against each other in a near-equilibrium that has the character not of a precisely definable pattern but rather a slightly fluid one." Organic form like the Beauty *sephirah* simultaneously exhibits balance and the lack of it, homeostasis and homeorhesis. Homeostasis is a balanced wholeness, integrity, and self-regulation. Homeorhesis is a fluid structure of growth, transformation and evolution. A cell's integrity is maintained by a semi-permeable membrane. The same membrane that contains the protoplasm is also an organic system that regulates what flows into and out of the cell. Although the cell is an integral unit, it can grow and reproduce itself. Encoded in its DNA, an individual cell contains both information for its own processes of homeostasis and homeorhesis, as well as the entire set of information for the balance and growth of the entire organism.

Like a cell, each *sephirah* contains all the *sephirot* and maintains its integrity while fluidly interacting with other *sephirot* through pathways of Hebrew letters. This organic analogy in the kabbalistic "Tree of Life" reflects all of Jewish consciousness. In his study of rabbinic thought, *Organic Thinking*, Rabbi Max Kadushin writes:

> The coherence of rabbinic theology can only be characterized as organic [...] The organic complex is flexible, allowing for divergences in interpretation either wide or subtle without losing thereby its essential unity. Fluid yet unified, the organic complex gives room for differences in temperament among individuals, even for different moods in the same individual, for stressing of different concepts in different historical periods [...] It

is organic coherence, not hard-and-fast logical consistency that permits the development and expression of individuality.[44]

The Beauty *sephirah* is the beating heart of the matrix of interconnecting pathways flowing to and from realms of intention, thoughts, emotions and action. Beauty is experienced as the innermost node in a dynamic network of connections and relations. Beauty is the aesthetic core of the creative process. In Jewish tradition as revealed through *kabbalah*, beauty is a vital innermost quality from which all emotions flow rather than attention to surface that characterizes definitions derived from Hellenistic aesthetic values. Kant's philosophical view that an "entirely disinterested satisfaction is called beautiful"[45] supports Bullough's psychological concept of "psychical distance"[46] in aesthetic experience. Prall writes in his book *Aesthetic Judgment*: "It is characteristic of aesthetic apprehension that the surface fully presented to the senses is the total object of apprehension. As we leave this surface in our attention, to go deeper into meanings and more broadly into connections and relations, we depart from the typically aesthetic attitude."[47] The Hellenistic aesthetic values expressed by Kant, Bullough, and Prall which limits the experience of beauty to apprehension of surface is diametrically opposed to the Hebraic values learned through *kabbalah* that Beauty is the dynamic vital inner core of connections and relationships linking the worlds of intention, thought, feelings and action. The Hellenistic and Hebraic attitudes towards beauty in aesthetic experience are so radically different that art forms created in response to these opposing sets of aesthetic values must be poles apart.

According to an anthropomorphic metaphor in ancient kabbalistic texts, Beauty is the pulsating heart energizing all the other parts of the body symbolized by the *sephirot*. The Hebrew letter pathways are the arteries and veins bringing the vital flow from the Beauty heart to the body's extremities. The Beauty heart's connection to the right arm of Compassion is through *h̲et*, to the left arm of Strength is through *tzadi*, to the right hip and leg of Success is through *yud*, and to the left hip and leg of Gracefulness is through *tet*. The right and left hemispheres of the brain, Wisdom and Understanding, are linked to Beauty through the pathways *samekh* and *ayin*. The Crown of the head is connected to Beauty through *dalet*. And Beauty connects to Foundation through *resh*. Foundation symbolizes the male genitals that bring the whole creative process to life though the pathway of the final letter of the alphabet *tav* flowing into Kingdom, the female genitals that receive, conceive, and give birth to a new being in the material world.

Postdigital aesthetic involvement with the actual living human body in action parallels this kabbalistic metaphor. Philosopher Richard Shusterman coined the word "somaesthetics" to describe the body "as a locus of sensory-aesthetic appreciation (*aistheis*) and creative self-fashioning." Unlike representational art that freezes the human body in paint or stone, postmodern somaesthetics "is not at all directed at representational beauty but instead at the quality of immediate experience: the endorphin-enhanced glow of high-level cardiovascular functioning, the slow savoring awareness of improved, deeper breathing, the tingling thrill of feeling into new parts of one's spine."[48]

Notes

1. Abraham Isaac Kook, *Abraham Isaac Kook: Lights of Holiness*, trans. Ben Zion Bokser (New York: The Classics of Western Spirituality. Paulist Press, 1978), p. 216.
2. Joseph B. Soloveitchik, *Halakhic Man*, trans. Lawrence Kaplan (Philadelphia: The Jewish Publication Society, 1983), p. 108.
3. Nosson Sherman (trans. and commentary), *The Complete ArtScroll Siddur* (Brooklyn, New York: Mesorah, 1987), p. xvi.
4. Alexenberg, "On the Creation of *LightsOROT*: Spiritual Dimensions of the Electronic Age," p. 55.
5. Nicholas Negroponte, *Being Digital* (NewYork: Alfred A. Knopf, 1995), p. 14.
6. Ibid., p. 12.
7. Lev Manovich, *The Language of New Media* (Cambridge, MA: MIT Press, 2001), p. 217.
8. Aryeh Kaplan, *SeferYetzirah:The Book of Creation* (York Beach, ME: Samuel Weiser, 1990).
9. Shelia Britton and Dan Collins (eds.), *The Eighth Day: The Transgenic Art of Eduardo Kac* (Tempe, AZ: Institute for Studies in the Arts, Arizona State University, 2003), p. 8.
10. Edward Lucie-Smith, "Eduardo Kac and Transgenic Art," *The Eighth Day: The Transgenic Art of Eduardo Kac*, p. 20.
11. Yitzchak Ginsburgh, "The Divine Emanations - The Ten Sefirot", Accessed October 7, 2010, (www.inner.org is a good website for seriously learning the basics of *kabbalah.*)
12. Zalman Schachter, *Fragments of a Future Scroll: Hassidism for the Here and Now* (Mount Airy, PA: B'nai Or Press, 1982), pp. 41–44
13. Soloveitchik, *Halakhic Man*, p. 101.
14. Ibid., p. 99.
15. Arthur Green, *Ehyeh: A Kabbalah for Tomorrow* (Woodstock, Vermont: Jewish Lights, 2003), p. 59.
16. M. Alexenberg, *Aesthetic Experience in Creative Process* (Ramat Gan, Israel: Bar-Ilan University Press, 1981).
17. A. Melnikoff, "Rabbi Kook on Art," *The Jewish Chronicle*, September 13, 1935.
18. Stephanie Barron, "The Galerie Fischer Auction," *"Degenerate Art": The Fate of the Avant-Garde in Nazi Germany*, Stephanie Barron (ed.) (Los Angeles: Los Angeles County Museum of Art, 1991), pp. 135–169.
19. *Midrash Pirkei d'Rabbi Eliezer* 36 from Lawrence Kushner, *The River of Light: Spirituality, Judaism, Consciousness* (Woodstock, VT: Jewish Lights, 1990).
20. Arthur Koestler, *The Act of Creation* (NewYork: Macmillan, 1964), p. 146.
21. Ori Z. Soltes, *Fixing the World: Jewish American Painters in the Twentieth Century* (Hanover, NH: University Press of New England and Brandeis University Press, 2003), p. 131.
22. AP, "Rembrandt Angel Flies Across Globe Via Fax," *Billings* (Montana) *Gazette*, October 6, 1989; AP, "Rembrandt Copies Via Fax", *Marion* (Ohio) *Star*; October 5, 1989; AP, "Rembrandt Drawing Meets Fax Age," *Shelby* (North Carolina) *Star*, October 5, 1989.
23. AT&T, *1989 Annual Report*.
24. Lippard, *Mixed Blessings: New Art in a Multicultural America*, (New York: Pantheon, 1990), p. 5.
25. Abraham H. Maslow, *Psychology of Science* (New York and London: Harper and Row, 1966), p. 137.
26. Albert Einstein, *The World As I See It* (New York: Philosophical Library, 1949), p. 5.
27. Ibid., p. 91.

28. Aryeh Kaplan, *The Living Torah* (New York and Jerusalem: Maznaim, 1981), *Exodus 3:2–14*, 269–271.
29. Donald A. Glaser, "Presentation of Award and Acceptance Speech, 1960 Nobel Prize for Physics," *TEID Library Services: Nobel Laureates*, http://www.lbl.gov/LBL-PID/Nobelists/LibD_Glaser.html. Accessed October 7, 2010.
30. Ibid.
31. Alexenberg, *Aesthetic Experience in Creative Process*, p. 72.
32. Ibid. p. 73.
33. Anton Ehrenzweig, *The Hidden Order of Art* (Berkeley and Los Angeles: University of California Press, 1967), p. 4.
34. Alexenberg, *Aesthetic Experience in Creative Process*, p. 156.
35. Ibid., pp. 197–202.
36. Ibid., p. 93, 94.
37. Ibid., p. 95.
38. Ibid., p. 109, 110.
39. Ibid., p. 165, 166.
40. Henri Poincare, *Science and Methods*, trans. Francis Maitland (New York: Dover, n.d.), p. 22.
41. C. P. Snow, "Appreciations in Science," *Science*, vol. 133 (January 27, 1961), pp. 256–259.
42. Chaim Brovender, "Towards *Ahavat Hashem*: Art and the Religious Experience," in Jeffrey Saks and Susan Handelman (eds.), *Wisdom From All My Teachers: Challenges and Initiatives in Contemporary Torah Education* (Jerusalem: ATID and Urim Publications, 2003).
43. Waddington, "The Character of Biological Form," in Lancelot Law White (ed.), *Aspects of Form* (Bloomington: Indiana University Press, 1968).
44. Max Kadushin, *Organic Thinking: A Study of Rabbinic Thought* (New York: Bloch, n.d.), pp. 12–14.
45. Kant, *Critique of Judgment*, p. 55.
46. Bullough, "Psychical Distance as a Factor in Art and an Aesthetic Principle," pp. 406–411.
47. David Wright Prall, *Aesthetic Judgment* (New York: Thomas Y. Crowell, 1929).
48. Shusterman, *Performing Live: Aesthetic Alternatives for the Ends of Art*, p. 137.

Chapter 5

Wiki Perspectives: *Multiform Unity and Global Tribes*

This may be the birth of a new era, perhaps even a golden one, on par with the Italian renaissance or the rise of Athenian democracy. Mass collaboration across borders, disciplines, and cultures is at once economical and enjoyable. We can peer produce an operating system, an encyclopedia, the media, a mutual fund, and even physical things like a motorcycle [...] Billions of connected individuals can now actively participate in innovation, wealth creation, and social development in ways we once only dreamed of. And when these masses of people collaborate they collectively can advance the arts, culture, science, education, government, and the economy.[1]

Don Tapscott and Anthony D. Williams

The divine purpose of the present information revolution, which gives an individual unprecedented power and opportunity, is to allow us to share knowledge – spiritual knowledge – with each other, empowering and unifying individuals everywhere. We need to use today's interactive technology not just for business or leisure but to interlink as people – to create a welcome environment for the interaction of our souls, our hearts, our visions.[2]

Menachem M. Schneerson

The earliest on-going wikicreation that honors multiform unity is the Talmud, a collaborative project of hundreds of people that spans centuries and continents. The oldest living global tribe is the Jewish People that has maintained its identity for millennia while being scattered across the globe. The postdigital age empowers networked people everywhere to participate in wikicreations and in forming new global tribes. A postdigital perspective is a Jewish perspective.

The Talmud's expression that there are seventy faces to every idea encourages creative, associative, open-ended, and alternative patterns of thought. The Talmud (which means "learning") is a prototypic open source wikicreation developed through collaboration, dialogue, arguments, commentaries, commentaries on commentaries, and multiple viewpoints. It can even accept completely opposite viewpoints as equally valid theoretically as in the difference of opinion about lighting Hanukah candles between Shamai and Hillel. Since the lighting of the candles commemorates the cruse of oil that lasted eight days in lighting the menorah during the rededication of the Temple in Jerusalem, Shamai proposed lighting eight candles on the first night of Hanukah removing one each night until there is one candle glowing on the last night. Hillel proposed the opposite procedure – lighting one candle on the first night adding another candle each subsequent night until all eight candles are burning on the last night. Shamai's proposal makes sense conceptually since all the oil was available on the first night, but Hillel's is more aesthetic and spiritual, brightening the world by adding more light. Jewish practice goes according to Hillel with the hope that in the messianic age when the world is so full of light, that Shamai's more conceptually valid way could be followed.

Like renowned Yale computer science professor David Gelernter who wrote ground-breaking books on both digital aesthetics and Judaism, Douglas Rushkoff is the author of *Nothing Sacred: The Truth about Judaism*[3] and ten best-selling books on new media such as *Media Virus*[4] and *Cyberia: Life in the Trenches of Hyperspace*.[5] He is a winner of the Marshall McLuhan Award for best media book and correspondent for the open source PBS project *Digital Nation: Life on the Virtual Frontier*[6] that explores what it means to be human in an entirely new postdigital world. In *Nothing Sacred*, Judaism is presented as an "open-source" tradition in which everyone has access and can participate. It is a tradition that stresses open-ended inquiry, transparency, and a commitment to conscious living. Millennia ago it developed media literacy by replacing hieroglyphics (literally "priestly language") with the aleph bet creating a tribe of literate people equipped to collaborate in responding to the challenges of changing times and places.

Members of this literate global tribe continue to participate in adding new interpretations (*h̲iddushai torah*) to the Jewish wikicreation through interactions locally with learning partners and globally through the Internet. Philosopher Martin Buber argues that it is not "a static unity of the uniform," that Jews aim for, "but the great dynamic unity of the multiform in which multiformity is formed into unity of character."[7] The multiform associative play with words and concepts that characterizes Jewish learning is appropriate for exploring the Hebrew word "*shevet*" and its English translation "tribe." According to *Wikipedia*, the word "tribe" first occurs in 13th-century Middle English literature as referring to one of the Twelve Tribes of Israel.[8] *Shevet* means the branch of a tree and a stick or rod used in weavings (related to networks/webs) and a tribe (related to family trees). As an expression of multiform unity, we read in the Bible how Jacob called his sons together to give each one a different blessing reflecting his unique mission as a leader of one of the twelve tribes (*Genesis 49:1–28*). In contemporary usage, the Hebrew expression "to undergo a stick (*shevet*) of criticism" refers to an artist being whacked by an art critic's comments about his artwork.

Checking Amazon, I found two books titled *Tribes*, one by Joel Kotkin and the second by Seth Godin. Kotkin[9] focuses on five groups: Jewish, British, Chinese, Japanese and Indian. In trying to explain the material success of these particular "tribes," he emphasizes historical patterns common to them all: a strong ethnic identity that allows the group to undergo economic and political changes without loss of essential unity; a global network based on mutual trust and communal self-help; and an open-minded approach to the adoption of scientific and technological innovations. Godin[10] defines a tribe as a group of people connected to one another, connected to a leader, and connected to an idea that inspires their passion. He argues that human beings have a need to belong, to be part of a tribe, to contribute to (and take from) a group of like-minded people. The Internet has eliminated the barriers of geography, cost, and time enabling countless new tribes to be born and mobilized around passions ranging from global warming to political campaigns. Saturday night, February 6, 2010, I created a new tribe "Postdigital Art" as a Facebook group with an accompanying "Postdigital Art" blog (http://postdigitalart.blogspot.com). A week later, my tribe had 248 members and had artists posting their artworks on the tribe's wall in response to my request for participation.

Some of the first websites listed in a Yahoo search for "tribes" are tribes.net, tribes.org, and tribes.com, three "Tribes" organizations dealing with online communities, art communities, and learning communities. Tribes.net is a free online community where members join existing tribes that were formed around common interests or form new ones to share discussions, pictures and reviews. Tribes.org is an arts and cultural organization that reflects and celebrates the fluidity and diversity of contemporary society. Tribes' audience comprises every possible ethnic group, age group, religion, and income level. The artists that tribes serve are similarly diverse. Tribes.org not only serves various communities on the Lower East Side of New York City, it actually creates a community. In this pan-disciplinary, multicultural environment, artists exchange ideas, create peer relationships, and find mentorships. Tribes.com develops learning communities in thousands of schools throughout the United States, Canada, Australia, and other countries that create safe and caring environments in which children at-risk can flourish. The purpose of the Tribes Learning Communities process is to assure the healthy development of every child so that each one has the knowledge, skills, and resiliency to be successful in a rapidly changing world.

The word "wiki" was appropriated by Ward Cunningham to refer to collaboration and peer production in Web-enabled communities, global tribes formed around common interests. He recalled a Honolulu International Airport counter employee telling him to take the quick Wiki Wiki shuttle bus that runs between the airport's terminals. He chose WikiWikiWeb in 1995 as an alliterative substitute for "quick" rather than calling his website Quick-Web. *The Wiki Way: Quick Collaboration on the Web*[11] describes the essence of the wiki concept as seeking to involve the visitor in an ongoing process of creation and collaboration that constantly changes the Web site landscape. Its most successful application is *Wikipedia*, a free online encyclopedia whose 14 million articles (3.1 million in English) were written collaboratively by volunteers around the world. There are more than 100 million article views per day. *Wikipedia* and its sister projects (*Wiktionary*, *Wikinews*, *Wikiquote*, *Wikisource*, *Wikispecies*, *Wikiversity* and *Wikibooks*), in more than 260 languages across the globe, are all being run by a non-profit foundation that has only a handful of employees.

"Wiki" has evolved new more generic meanings in multiple fields when activities involve collaboration, participation, peer production, collective creativity, and joint projects. Telling the the story of how collaborative virtual prospecting saved a failing business can illuminate options for wikiart projects. Tapscott and Williams begin their book *Wikinomics: How Mass Collaboration Changes Everything*[12] with a case study of the Canadian gold-mining firm Goldcorp on the verge of bankruptcy that was taken over by a young mutual fund manger with no experience whatsoever in the field. The company's gold mine was dying without evidence of new gold deposits anywhere on Goldcorp's property. The company's geologists made an unsuccessful last bid attempt to find gold to prevent the closure of the company. The new Goldcorp CEO had the epiphany while at an MIT seminar that if his employees could not find gold, maybe others could. In a secretive industry in which geological data is carefully guarded, he made all the company data public on the Internet launching a contest with cash prizes to participants who came up with the best methods of finding gold on the company property. As news of the contest spread around the Internet, entries flooded in from more than one thousand virtual prospectors from fifty countries, not only from the field of geology, but from computer graphics, mathematics, physics, intelligent systems, the biological sciences, and other fields. Some of the highly original ideas that emerged from the wide pool of contestants resulted in finding copious quantities of gold that catapulted a business on the verge of collapse into a nine billion dollar enterprise.

Desert earth and terrestrial isopod tribes

I initiated a protowiki participatory artwork in the summer of 1980 when I took a break from my work as head of a college in Israel's Negev desert to be thrust into the center of the high tech world as a research fellow at MIT's Center for Advanced Visual Studies. In addition to my work developing a biofeedback-generated interactive visual imaging system, I created the *International Desert Earth Archives* (*I.D.E.A.*), a pre-digital participatory artwork that invited the collaboration of people from the 44 countries that have deserts. This artwork was inspired by the commentary on *Genesis* that Adam was formed from earth collected from places throughout the world so that no one nation could claim to be the original descendent from Adam. Perhaps the desert *MiDBaR* would speak *MiDaBer* of world peace. (The Hebrew words for desert and speak are written with the same four letters *MDBR*). This project became a precursor to my other wikiart projects with local communities and global tribes.

I sent a letter (in the days before e-mails) from the International Desert Earth Archives on the letterhead of the Massachusetts Institute of Technology to the each of the embassies in Washington of countries with deserts. My letter, addressed to the ambassador, invited his country to participate in the International Desert Earth Archives by sending me a sample of desert earth, a photograph of the collection site, and a map marking the site. I signed the letter "Professor Mel Alexenberg, Director of I.D.E.A."

My artwork presented the photographs, maps, earth samples, the containers in which the earth was sent, and documentation of the political process that revealed the great difficulty government bureaucracies have in figuring out whose role it is to collect earth. In response to my request, I generally received a letter back for each country that was signed by the embassy's scientific or economic attaché indicating interest in participating. No country asked why I wanted the earth. Then an amusing paper trail followed. The embassy forwarded my request to their foreign ministry that forwarded it to the ministry of the interior, science, or industry that forwarded it to the ministry of agriculture that then sent a letter back to the foreign ministry stating that they had no manpower to go out into the desert to scoop up some sand. Sometimes a more scientifically oriented agriculture ministry would send an inquiry about what kind of earth I needed. After months, some embassies would phone me to tell me that their country could not participate since they could not find anyone in their government who could collect the earth. When I responded incredulously, "What? You don't want your country represented in the International Desert Earth Archives?" they invariably said that they would try again. I then collected documention of a repeat performance.

Argentina was the first country to send me desert earth. Six months had past since I sent my request. It was delivered by diplomatic courier in two cartons closed with official wax seals along with photographs and a hand-drafted map on velum. On average, it took three years until countries could figure out how to collect earth. Since I had not specified how much earth I wanted, I got different quantities in interesting kinds of containers from the two huge Argentinian cartons to a 35mm film canister from Botswana to a white cloth sack sewed closed covered with Arabic script from Kuwait. My photographs of the containers became part of the artwork when I exhibited *I.D.E.A.* in the 1988 exhibition *Golem! Danger, Deliverance and Art* at the Jewish Museum in New York.

Although I had received earth from Asia, Africa, Australia, and South America, I had not received any from Europe because it is a desertless continent and from North America because the United States has no embassy in Washington. In order to get North American desert earth, I sent the same letter that I had sent to embassies in Washington to the ambassador of the United States in Tel Aviv on a letterhead from Ramat Hanegev College rather than on an MIT letterhead. I received a response from Samuel Lewis, the ambassador himself, that he had passed my request on to Dennis Jett, the scientific attaché at his embassy. Handing my request over to the scientific attaché seemed to be the beginning of the usual protracted bureaucratic process. However, three weeks later, I received a phone call at my office in the Negev from Dennis Jett who informed me that he had earth for me from the desert in New Mexico. He drove down to my college to hand deliver to me a Hills Brothers coffee tin filled with earth from White Sands, the site of the world's first atomic bomb detonation, along with the collection site marked on a US Geological Survey map and a large color photograph of the rocky desert terrain. I asked him how he was able to obtain all I had requested in three weeks when it took years for other countries to comply. He explained that if he followed the official procedure of sending my request to the US State Department it would have taken years, too. What he did was phone a college friend from the University of New Mexico who had become a geologist working at the research center in White Sands and asked him to send me what I wanted. His friend wanted to know why I needed more desert earth living in the middle of the Negev desert. Jett took the two-hour drive from Tel Aviv to see me and find out. Ambassador Lewis came to the dedication of a new arts building at my college where he saw my first exhibition of *I.D.E.A.* in which I had incorporated his letter into my documentation of this international participatory artwork. Perhaps America's great success growing from a small British colony to becoming the leader of the free world is based on honoring creative thinking and the spirit of short-circuiting governmental red tape. Checking *Wikipedia* to confirm that Jett is spelled with two t's, I was happy to learn that he was rewarded for his creative problem-solving ability by subsequently being appointed US Ambassador to Peru and Mozambique and professor of international relations at major American universities.

Fast forward from my participatory pre-digital artwork to the postdigital world of collaborative networking that spawned YASMIN, a moderated list for art-science-technology interactions around the Mediterranean Rim. In 2008, YASMIN launched a discussion "Invisible Dynamics: Mediterranean Diasporas, Networks, Processes." The discussion moderator described the Mediterranean region as a complex of cultural, ecological, environmental processes that shape its human, ecological, and physical history and evolution. Many of these processes remain part of an invisible dynamics below our awareness. Yasminers were invited to explore ways that human, animal and plant migration and diasporas shape human perception and create artworks dealing with the dynamics at work around the Mediterranean Rim.

The YASMIN discussion inspired me to create *MERTI: MEditerranean Rim Terrestrial Isopods*, a wikiart project that invites collaboration in exploring the biogeography of terrestrial isopods as an aesthetic metaphor for human habitation in the 21 modern states that have a coastline on the Mediterranean Sea. Terrestrial isopods, commonly known as sowbugs, pillbugs or woodlice, are crustaceans more closely akin to shrimps and crayfish than to insects. They are the only crustaceans that have adapted to living their entire life on land. You can probably find these little creatures in your own back yard under woodpiles, stones, and flower pots, in leaf litter, and in other damp microhabitats. Most isopods live under the sea except for the small gray creatures in the suborder Oniscidea that have evolved into hundreds of species that live on land but breathe with gills. What is most relevant to the YASMIN discussion statement is the fact that each of the 21 Med Rim countries has human indigenous and migratory Med Rim populations that find parallels in the populations of native and introduced species of global terrestrial isopod tribes. *Armadillidium vulgare* is an isopod species that forms a global tribe that has established colonies in all the Med Rim countries. It is commonly called a pillbug and affectionately called a roly-poly by children because it can roll up into a tight ball to hide its gills and present the minimal surface area to minimize water loss. I know these pillbugs well from my childhood in New York and my research as a biologist. These cosmopolitan isopods have settled in port cities worldwide. Canadian biology professor Thomas Carefoot wrote in his paper "Toward the New (Isopod) Millennium" that "they did this by 'riding the coat-tails' of settlers, infesting outward-bound luggage and cargo, and perhaps by floating about the oceans on debris."[13]

I posted on YASMIN an invitation to collaborate in my *MERTI* wikiart project by helping to collect data on the different species of terrestrial isopods around the Med Rim. This data-collecting process forms an integral part of the artwork that relates the ecology, evolution, and biogeography of native and introduced species to patterns of human migrations and diasporas. I asked my fellow Yasminers to search for pillbugs in their local environments and photograph or draw them for identification by their university biology departments and research laboratories.

This wikiart project that invites the collaboration of Yasminers and Med Rim biologists renews a biogeography project that I began during 1957–58 when I wrote my senior honors thesis on the comparative ecology of the terrestrial isopods *Armadillidium vulgare* and *Oniscus asellus* at Queens College, City University of New York, under the supervision of the renowned biologist Max Hecht, editor of *Evolutionary Biology*.[14] Unlike the rapid dissemination of my request for collaboration in a networked world, I collected data on the global distribution of these two species via snail mail. As biologists worldwide slowly mailed me data, evidence mounted up about the congruence between the biogeographic distribution of these two species of sowbugs and of human populations. The sowbugs that I studied under decaying wood on a Long Island lot may have found their way there from a port city on the northern rim of the Mediterranean Sea through the port of New York under wooden shipping crates.

My lingering affection for these amazing little creatures surfaced periodically during the past half-century from my 1958 honors thesis to the 1962 publication of my taxonomic key to the sowbugs of Long Island and my article on pillbugs for children in the magazine of the American Museum of Natural History,[15] my paintings of their interactions, my photographs documenting terrestrial isopods at the four corners of America (Miami, San Diego, Seattle, and Portland, Maine) in 1996, to my wife creating a huge ceramic pillbug for me on my 60th birthday in Miami when we working on our *Legacy Thrones* intergenerational collaborative artwork.

Legacy Thrones: intergenerational collaboration

Through aesthetic dialogue between young people and elders from Miami's three major ethnic tribes – Hispanic, African-American, Jewish – valued traditions were transformed into artistic statements of enduring significance. Together, young and old hands shaped wet clay into colorful ceramic relief elements collaged onto three towering thrones constructed from steel and concrete. Each twenty-foot high, two-ton throne installed in a park facing Biscayne Bay preserves the cultural integrity of each of the three largest ethnic communities of elders who settled in Miami.[16] This environmental public art project is a realization of Gablik's assertion that postmodernism is creating "a new paradigm of *participation* in which art will begin to redefine itself in terms of social relatedness [...] encouraging the emergence of a more participatory, social interactive framework for art, and supporting the transition from art-for-art's-sake assumptions of late modernism."[17]

Creative teams of three, one elder and two art students, worked together one day each week for a full academic year. The students worked on this project within the framework of a course I taught on environmental public art at New World School of the Arts in Miami.[18] My artistic collaborator and wife, Miriam Benjamin, arranged for African-American elders to be bussed to the school from Greater Bethel AME Church, Hispanic elders from Southwest Social Services Program, and Jewish elders from the Miami Jewish Home for the Aged. All of those who volunteered to participate were women between 70 and 90 years old. This collaboration between artists, elders, and young people in creating a monumental work of public art enriches their shared environment and leaves a legacy for generations to come. In *The Third Hand: Collaboration in Art from Conceptualism to Postmodernism*, Charles Green proposes "that collaboration was the crucial element in the transition from modernist to postmodernist art and that a trajectory consisting of a series of collaborations emerges clearly from the late 1960's conceptualism onwards [...] Redefinitions of art and of artistic collaboration intersected at this time."[19]

The sixty participants worked simultaneously in one huge studio space with the artistic guidance of Miriam Benjamin and me. Collaborative art changes the visual artist's role to be more like a creative leadership role in the performing arts. Instead of the solitary role alone in one's studio, the postmodern paradigm finds the visual artist acting more like a choreographer in dance, composer/conductor in music, playwright/producer/director in theater and film.

At their first meeting, each pair of students listened to an elder tell about her life experience and cultural roots. Life review methodologies developed by Susan Perlstein at Elders Share the Arts in New York facilitated elders looking back and reaching inward to trigger reminiscences of events and images of personal and communal significance.[20] The challenge at the next meetings was to explore ways of transforming reminiscences that reveal cultural values into visual images that can be expressed through clay. The enthusiasm of the elders in being artistic collaborators in creating a monumental artwork is expressed by the eminent psychologist Erik Erikson: "For the ageing, participation in expressions of artistic form can be a welcome source of vital involvement and exhilaration [...] When young people are also involved, the change in the mood of the elders can be unmistakably vitalizing."[21]

Although nearly all of the elders had no prior experience in making art or working with clay, they developed their technical prowess and aesthetic judgment during their year of participation. While the students facilitated the elders' growth artistically, the young people's lives were enriched through creative collaboration with partners blessed with a long life of fertile experiences. By sharing their stories with the students, transforming them into artistic images, and leaving a legacy for future generations, the elders added deeper layers of meaning to their lives.

With the assistance of art students experienced in ceramics, the elders worked with clay to make relief sculptural statements depicting personally significant images from their rich life experiences. A Jewish woman who was a dancer in her youth with the Bolshoi Ballet in Moscow formed women dancing the *horah*, a traditional Jewish folkdance. An African-American woman made a mule-drawn wagon on which she rode to church as a child in rural Florida. A Cuban woman made high-heeled shoes and an elegant pocketbook, the only valued possessions she took with her while escaping from Cuba on a rickety boat that sailed across the Florida straits. In *Art Education: Content and Practices in a Postmodern*

Era,[22] Jones identifies postmodern art as art based upon personal and emotional metaphors and the acknowledgement of the importance of narrative and personal myth. Anderson writes in the same book: "In a postmodern framework, art is once again something beyond itself, it defines a particular narrative or world view."[23] That postmodern art is based upon collecting "little narratives," each presenting an alternative way of experiencing the world, is argued in *Postmodern Art Education.*[24] Little narratives show that each cultural story is but one among many.

Complimenting their personal images, the elders made images representing communal experiences and symbols of shared culture values. The Jewish group formed Hebrew letters, a Hanukah menorah, the biblical dove of peace, and symbols of the ten *sephirot* representing the stages in the parallel process of human creativity and divine creation. The African-Americans created images of black slaves in agony, cotton fields of the rural South, the keyboard of their church organ, and traditional African masks and geometric motifs. The Hispanic elders made a guitar and maracas, a cup of Cuban coffee, baseball players, fighting cocks, an Aztec bird, a rainforest frog, Jesus with outstretched arms, and Mary with a sunburst halo. The National Art Education Association book *Postmodern Art Education* defines art as "a form of cultural production whose point and purpose is to construct symbols of shared reality" and the value of art as promoting "deeper understandings of the social and cultural landscape."[25]

After the clay dried and was fired, the elders and young people painted them with colorful glazes. These relief ceramic forms became collage elements for covering the concrete surfaces of thrones until they were entirely clad in ceramics. Postmodern art media emphasize appropriation, collage, montage, pastiche, and juxtaposition of meaning from diverse images.

All three thrones were made the same size and basic shape with the form of each throne's crown and sides being different. This presented the semiotic statement that all three cultures were equal in status yet each was a unique expression of a different culture. The Hispanic throne has a sunburst crown and water waves cascading down the two sides. The Jewish throne is topped by a Hanukah menorah that holds nine flaming torches with an aluminum enlargement of leather straps meandering down the sides from a box containing biblical passages (*tefilin*) that is worn by Jews on their heads during morning prayers. On the head of the African-American throne is a giant African mask with its sides designed with a geometric pattern derived from a traditional African motif. The frameworks for each of the three thrones was constructed by welding steel pipes connected to each other with rebar rods to reinforce the concrete that filled the spaces between the pipes. Victor Arias, one of my art students who is a professional metal worker from Ecuador, built the steel framework. In order to move the two-ton thrones, wheels were welded onto them. The students worked on cementing the collage elements to the thrones while they rested horizontally. The elders did not participate in this part of the process since it was too physically demanding and time-consuming.

After the front and sides of the thrones were fully clad in ceramics, the two-ton thrones were transported to the park site by a large tow truck, lifted up by a crane, and installed at their permanent sites along the shoreline walkway of Margaret Pace Park. Standing tall made the unfinished rear of the throne backs accessible. Two of the students who had worked on the thrones when they had rested horizontally in the studio, took the responsibility for completing the three throne backs on site. Decorative Hebrew letters for the rear of the Jewish throne created by my art students at Emuna College in Jerusalem were shipped to Miami. Ceramic elements made by students and Jewish elders in Miami collaged together with those made in the Land of Israel forges a powerful link with the elders' ancestral homeland.

The *Legacy Throne* project spanned six years from the time the elders first came to work on creating them to sitting on the finished thrones with their grandchildren watching boats pass by in the bay. Each throne honors the integrity of a monoculture rather than a multicultural mix. Harold Pearse suggests that in postmodern art "what was heralded as multiculturalism will give way to polyculturalism."[26] Instead of being an expression of the outdated American concept of the "monotone meltdown pot," the thrones are more like vegetables in a salad, each retaining its own form and flavor. Lucy Lippard appreciates that "this model is fresher and healthier; the colors varied; the taste often unfamiliar."[27] In the book *Taking Part: A Workshop Approach to Collective Creativity*, Lawrence Halprin writes: "In a complex society such as ours which wishes to allow group differences to *emerge*, not *submerge*, we need to find ways for these groups to express themselves and be heard and valued. One of the major purposes of participation is to allow diversity to be expressed."[28]

On the other hand, although vegetables in a salad retain their individual flavors they taste better eaten together with a salad dressing. Working on the separate thrones, not only did they learn to appreciate their differences, the elders also realized how much they shared. The three different cultural groups continually engaged in dialogue with each other while working in one large studio, an opportunity that rarely exists in their everyday lives. African-American, Hispanic, and Jewish old women in their ethnically specific homes for the aged and senior centers seldom encounter each other. As they learned about each other's cultures, they came to realize how much they shared a common set of values. They jointly selected a theme for *Legacy Thrones* from the biblical passage: "Behold, how good and how pleasant it is when we sit together" (*Psalm 133*).

All three groups of elders shared their commitments to living in freedom and to biblical values. Gaining freedom from slavery and freedom from tyrannical regimes shaped their reminiscences. Some women had heard first-hand stories of slavery on Southern plantations from their grandmothers. Others were survivors of the Holocaust. One survivor spoke about having to bite the umbilical cord of her child born in hiding in an underground pit. Cuban exiles talked about escaping the brutal oppression on the island they loved.

Although the African-Americans were Protestant Christians, the Hispanics Catholics, and the Jews Jewish, they all shared an appreciation for the freedom in America that they recognized as germinating from a common set of biblical values. The sculptured images they created showed their shared attachment to the Bible. A ceramic book with the relief words

"Holy Bible" graces the African-American throne. All the African-Americans elders were members of a senior group at their church named Bethel, "House of God" in biblical Hebrew. The *sephirot* images and Hebrew letters on the Jewish throne are derived from the biblical books of *Genesis, Exodus*, and *Chronicles.* And biblical figures shaped in clay create a spiritual presence on the Hispanic throne. In *Mixed Blessings: New Art in Multicultural America,* Lucy Lippard summarizes the postmodern values of *Legacy Thrones*: "I am interested in cultural dissimilarities and the light they shed on fundamental human similarities" as well as "art that combines pride in roots with an explorer's view of the world as shared by others."[29]

Wikiart blog: Med Rim intercultural interaction

The *Wikiartists* blog was created in 2009 to invite people in all walks of life to become wikiartists by participating in the creation of web-enabled peer-produced artworks. This blog's URL is http://www.wikiartists.us – not .com, not .org, not .net – but .us – us – we the people interacting with each other through the Web, crossing borders and honoring cultures different from our own.

The first wikiart project featured on the *Wikiartists* blog is *MERIWIP: MEditerranean RIm WIkiart Project* inspired by YASMIN's call for projects to advance communication, collaboration, and cooperation between people living around the Mediterranean Sea. Participants from the 21 Med Rim countries (Spain, France, Monaco, Malta, Italy, Slovenia, Croatia, Bosnia-Herzegovina, Montenegro, Albania, Greece, Turkey, Syria, Cyprus, Lebanon, Israel, Egypt, Libya, Tunisia, Algeria and Morocco) are submitting pictures of their wedding celebrations. Other Med Rim couples send them native flowers as wedding gifts. Both the wedding pictures and flowers are posted on the blogart site in the geographical order of the countries as they surround the Mediterranean.

MERIWIP is creating a dynamic interchange between people living in the Med Rim countries promoting goodwill and unity – not the unity of the uniform, but multiform unity in which the distinctive culture of each country is honored. Marriage is an archetype of multiform unity, a loving unification that respects diversity. At its best, it creates a vibrant harmony between two singular people of different sexes, life experiences, cognitive styles and viewpoints. It is a model for bringing intercultural dialogue and peaceful relationships between the rich variety of peoples and nations surrounding the Mediterranean.

I asked people living in the Med Rim countries to translate into their language my invitation to become a wikiartist by participating in *MERIWIP*. I posted the translations as I received them. Of all the Med Rim languages, only Hebrew and Greek are still in use after more than 3,000 years. All the other languages are of more recent origins. Greeks write today with the words of their ancient bards. Israeli Jews can read the Bible in the original Hebrew with the same fluency and understanding as reading Hebrew-language Internet news sites.

I posted the following directions on the *Wikiartists* blog:

PEER PARTICIPATION: WEDDINGS AND FLOWERS

Step 1. Send a wedding photograph from your country, the first names of the bride and groom, and the city and country in which they live. E-mail the picture as a jpg file and the information about the couple to art@wikiartists.us. The wedding pictures will be posted on the *Wikiartists* blog in the geographical order of the countries as they surround the Sea.

Step 2. Send a photograph of flowers that grow in your country as a wedding gift to a couple from another Med Rim country. E-mail the flowers as a jpg file to art@wikiartists.us. Your flower picture with your name and the names of your country and flower species will be posted with the wedding picture of the couple receiving them.

New images and texts are continually being received. Francesca and Remo sent campanella flowers from Italy's Capri Island to Roka and Maija in Vodice, Croatia. Albireo sent wild flowers growing in a field near Mellieha in Malta to Jihan and Sari in Beirut, Lebanon. Samuel of Navara, Spain, sent an Itsas brabelina flower growing in the Basque countryside to Aisling and Barry in Slovenia. And Pnina sent white beauhinia flowers from Israel to Paul and Jo of Neja, Spain.

Unlike finished artworks hanging on museum walls, wikiart exhibited in cyberspace never needs to be finished. *MERIWIP* is an ongoing wikiart project through which people living around the Mediterranean Rim are continually invited to be wikiartists by sending their wedding pictures and local flowers for posting on the *Wikiartists* blog. As a reader of this book, you can collaborate in creating this artwork. If you live in a Med Rim country, send your wedding picture or the wedding pictures of your family and friends and photographs of flowers native to your country. If you live elsewhere, invite your Facebook friends, friends in Second Life, and real-space friends from the 21 Med Rim countries to participate.

JerUSAlem-USA: international partnership

JerUSAlem-USA is a participatory art project that links the twenty places in the United States called "Jerusalem" with the original Jerusalem in Israel from which they are named. There are Jerusalems in Alabama, Arkansas, Georgia, Maryland, Michigan, New York, North Carolina, Ohio, Rhode Island, Tennessee, Utah and Vermont. The name of these American Jerusalems was inspired by the Bible where we read about King David establishing Jerusalem as the capital of the unified state of the Israelite nation more than 3,000 years ago. Today, Jerusalem is the lively capital of the modern state of Israel.

I created a *JerUSAlem-USA* blog and a *JerUSAlem-USA* Facebook group where I posted invitations to participate in this art project by sending photographs of everyday life (people, homes, stores, community events and celebrations, flora and fauna, scenery, and signs showing the name "Jerusalem"). I also invited participation by sending letters to the editors of local newspapers read by people living in the American Jerusalems and neighboring communities. I also e-mailed my request to professors at art schools and college and university art departments in the dozen states with Jerusalems to invite their photography students to participate. I contacted governors, senators, congressmen, and state and local governments with requests to arrange collaboration by their constituents. I then posted on the project blog an invitation in Hebrew inviting Israelis to study the photographs of Jerusalems in the USA and submit matching images of Jerusalem in Israel.

The juxtaposition of photographs of the original Jerusalem with those from the twenty American Jerusalems creates an interactive network of people with shared values that deepens the friendship between them. A photograph of a wooden covered bridge built over the Jericho River by Thomas Forsyth in 1865 in Jerusalem Mills, Maryland, is juxtaposed with a photograph of the light rail bridge designed by Santiago Calatrava and built in 2009 to span the entrance to Jerusalem, Israel, drawing up images of the stringed harp of King David. The image of an Independence Day celebration in Jerusalem in upstate New York on the shores of Lake Keuka to commemorate the liberation of the USA from British rule on the 4th of July 1776 is matched by an image of a Jerusalem Day celebration in downtown Jerusalem commemorating the liberation of Jerusalem after nearly 2,000 years of occupation by foreign powers from the Roman Empire to the Ottoman Empire to the British Empire. A photo of Ave Maria Grotto, known as "Jerusalem in Miniature," a landscaped park designed to provide a natural setting for the 125 architectural miniatures created of stone and concrete as the lifetime work of Brother Joseph Zoettl, a Benedictine monk of St. Bernard Abbey in Alabama is posted with a photo of "Mini Israel," a tourist attraction midway between Jerusalem and Tel Aviv that features over 350 beautifully crafted exact-replica models of historical, religious, archeological, and modern sites in Jerusalem and throughout the Land of Israel. The fun photograph of a pink rabbit on a spring for children to ride in a playground in Jerusalem, Ohio, is juxtaposed with a photograph of a huge fantasy monster created by artist Niki de Saint Phalle in 1972 in Jerusalem, Israel; children climb up inside the monster to slide down any one of its three tongues.

People began to not only send me photographs of Jerusalems in the USA, but attached delightful stories of how the American Jerusalems got their names and descriptions of the places they chose to photograph. Residents wrote about their experiences living in an American Jerusalem and visitors wrote about their observations of the place and their discussions with people who live there.

Judi and Al Grant of Jerusalem, Rhode Island wrote: "My husband Al and I live in the village of Jerusalem year round. We heard about your project through an article written in our local paper – *The South County Independent*. The photos of the Block Island ferry were taken from our home looking over to the Port of Galilee. Rhode Island has four seasons and whether it's winter, spring, summer or fall each season brings a sense of contentment and well being within our community. It is truly a remarkable place. I've attached another photo – it was on the outside of our home when we purchased it years ago. The man that built the house in 1945 placed it on the chimney – Now the way to get to Jerusalem is by way of

Galilee passing through Jericho – is it not? Was this the original owner's way of saying this was his little slice of heaven?" About how it got its name. "In 1902, the story goes, Thomas Mann a fisherman from Nova Scotia who had settled here, felt the village that had sprung up with its fishing shacks should be called Galilee after the fishing village of biblical times. One day, an old timer sat on the docks repairing his nets when a stranger called out to him, 'Where am I?' The answer was 'Galilee.' 'And what is that?' the stranger asked pointing to the other side of the channel. The old timer thought for a minute, nodded his head and replied, 'Must be Jerusalem!' And so the names of Galilee and Jerusalem have been used since to denote a most picturesque part of Rhode Island."

Molly Davis, a student of photography professor Amy Jorgensen at Snow College in Ephraim, Utah, wrote: "A long, dry, dirt road led up to the town of Jerusalem, Utah. There were old decaying cabins and sheds, log cabins, and tiny houses where the people lived. A car came down the dirt road with a lady wearing a pink hat and pink lipstick. She introduced herself as Laura Phelps, 'Mayor' of the town. She was sweet and excited about the JerUSAlem-USA project when we explained it to her. She told us about the history of little Jerusalem that was established by her great-grandfather. Walking down the road was a man in his flannel shirt and worn out jeans with his eight-month-old Great Dane puppy. You could hardly call her a puppy. She was huge. The man was in the process of training her. Her jumping all over the place scared me. The man, who did not want to be identified, told us about his great-great-grandmother and how she lived in the broken down blue bus in his shed. He then left to help his neighbor with a farming project. The farming equipment was painted in beautiful bright colors, which was humorous because the town itself was very dry and bland with green trees and hills behind it. I felt a sense of beauty there even with the 'No Trespassing' and 'Keep Out' signs."

Cynthia Glad is also a student at Snow College where she is majoring in art. She is a 38-year-old mother of six who interviewed and photographed three of the families living in Jerusalem, Utah. Cynthia lives in the town of Moroni only a five-minute drive from Jerusalem. Just as different photographers of the same Jerusalem focus on different aspects of the place, the texts I received complement each other by providing alternative viewpoints. Cynthia wrote: "I found as I talked to the residents of Jerusalem, they were excited as well. I think this gives one more human connection to the rest of the world that sometimes these little places are lacking. They are happy and content in their lives, secluded for the most part by choice. But to know that there are people just like them in communities across the USA and in Israel is eye opening. Even though the initial connection is linked only by the name of their community, they will find other similarities with the people and lives of those other places. It makes the world a little smaller. Kind of like Disney's 'It's a Small World After All.' One couple spent many years taking their children to the farm on weekends and holidays. When the father retired and sold his chiropractic business, they went to the farm to live. They own several acres and have a gigantic garden. There are bags and boxes of fruits and vegetables throughout the kitchen, as well as bottles and racks of dried and processed foods. The night I visited with them, the pressure canner was busy at work on the stove. All these are signs of the efforts they have put into the land and the reward for their hard work. They have one pasture that every evening fills up with about a hundred deer. They lead a busy life that at the same time carries a sense of quiet. The land that Jerusalem, Utah, sits on has been in the mother's family since the mid-1800s when the first settler Lawrence Christensen moved there. Laura Phelps, another great-granddaughter, says of her great-grandfather, 'Someone asked him the name of his little town and he said, Little Jerusalem. It has been Jerusalem ever since.' Jack Bailey drove by in his truck. He is the great-grandson of Lawrence Christensen. He was born and raised in Jerusalem. His family home had burned down many years ago. He now has a home that sits on many acres and is nestled against the mountain where he raised his kids. Jack is 82 years old and has been farming all his life. Jack has only about 30 head of sheep left on his farm. He said, 'I'm about ready to give it up.'"

***Photograph God*: *kabbalah* through a creative lens**

Scot Kaplan and Sarah Weinstock, curators of the Wayne and Geraldine Kuhn Fine Arts Gallery at the Marion Campus of the Ohio State University, invited me to show my interactive artblog *Photograph God* in the *Searching for God* exhibition in 2010. In the 20th century, when my paintings and prints were in exhibitions, I had to arrange to transport them there by truck, train or plane. Since it was too expensive to transport and insure my large acrylic painting from Tel Aviv to Prague for my 2004 *Cyberangels: Aesthetic Peace Plan for the Middle East* exhibition, I e-mailed the Prague museum curator a photograph of it that was then digitally printed out on canvas in Prague at the same size as the original. For the Ohio State University exhibition, I did not need to ship the artwork from Israel to Ohio nor even e-mail it. Since my *Photograph God* artblog had been living and growing in cyberspace for four years and can be accessed by anyone from anywhere, the exhibition curator only had to google "photograph god" to find it the first of a list of over six million sites. One click of the mouse and it was there for the taking.

This experience brought to mind Walter Benjamin's seminal essay on art in the age of mechanical reproduction that explores the relationships between original and copy that I discuss in the "Semiotic Perspective" chapter in relation to the "Made by God" original Ten Commandants that Moses smashed to bits. In the postdigital world, the original is written in invisible bits and bytes. There are neither originals nor copies. The invisible code can be played out in alternative ways from large-screen projection to a computer monitor to the iPod screen. Boris Groys in his book *Art Power*[30] proposes that the digital image that has no visible original is not merely exhibited but staged, transforming the visual arts into a performing art. Performing the invisible image is analogous to performing a silent musical score. Like the silent score that can only be heard when performed by musicians, the invisible digital code can only be seen when performed by curators. The role of the curator becomes more powerful in the postdigital era as she not only becomes an exhibitor but the performer of the image. In addition to the traditional curatorial role of hanging pictures on the wall in relation to each other and the gallery space, the contemporary curator turns the invisible into the visible. When curators Scot Kaplan and Sarah Weinstock showed my *Photograph God* blog in the Ohio State University exhibition, they became my partners in creation as musicians are a composer's partners in bringing the artform to life. They assumed control of what my artwork would look like by deciding how to play it out in the gallery, whether on a small computer monitor on a table with a mouse for individual users, on a large flat-screen television screen mounted on the wall with automated control, or a full-wall projection in a separate room. Kaplan proposed the concept of "the fifth wall" in response to my request to learn how my work was being shown. In his e-mail to me, he wrote "Your piece is available for view on a computer terminal located in the gallery. This is also our first experiment with what we are calling 'the fifth wall' which is a series of links to works that we have included in the show but whose primary content exists on the Web which we see as a virtual fifth wall of our traditional 'white box' or 'cube' gallery."

The postdigital interplay between invisible and visible realms is further explored in Sonvilla-Weiss' book *(In)Visible: Learning to Act in the Metaverse.*[31] He relates visibility/invisibility to the highly charged dichotomies of private/public, individual/community, autonomy/control that engage the cognitive, social, technological, and aesthetic dimensions of a dataculture that is attempting to mark new routes to communicate strategies and practices of artistic engagement. The stage was set for electronic routes to render the invisible visible in the pre-digital world of analog technologies. My 1988 editorial as art editor of *The Visual Computer: International Journal of Computer Graphics* that I titled "Art with Computers: The Human Spirit and the Electronic Revolution"[32] took a prescient look at postdigital perspectives that humanize technologies by revealing invisible realms that give us clues to the worlds that kabbalists described centuries ago. I wrote how the electronic age of extended mind and global range carries within it seeds of spiritual renewal. Consider that in the very room in which you are reading this book thousands of events from throughout the world are happening simultaneously: a concert in Moscow, a baseball game in Los Angeles, cooking lessons in Jerusalem, a carnival in Rio, ski jumping in the Swiss Alps, and a bicycle race in the south of France. You may ask, "What are you talking about? My room is quiet and empty. The only event occurring in my room is my act of reading this book." Think, however, that when you turn on your TV you can see all these events that have been silently present in your room all the time. These events have been transformed into patterns of electromagnetic energy that cannot be perceived by our ordinary senses. Invisible, they permeate our environment even passing unnoticed through our bodies. Turn on the TV, however, and these hidden realms are revealed in full color.

I created my blogart site *Photograph God* (http://www.photographgod.com) in 2006 as a project for my students at Ariel University and Emuna College in Israel. Since a blog can be accessed worldwide, people in other countries began to participate by submitting images of their experiences. I posted on the blog an invitation for Ohio State students and visitors to the *Searching for God* exhibition to participate in the *Photograph God* wikiart project by e-mailing me their photographs and explanatory text. My invitation was followed by instructions on the blog on how to make an invisible God visible:

FOCUS YOUR LENS ON GOD

Focus your camera lens on God and you will see God look back at you. Seeing God is seeing divine light reflected from every facet of your life. The ancient wisdom of *kabbalah* will help you recognize that you have been looking at God all the time but missed the action. The Bible invites you to find God in "the Compassion, the Strength, the Beauty, the Success, the Splendor, and the Foundation of everything in heaven and earth." (*Chronicles 1, 29:11*)

You only see light. You have never seen your mother, father, spouse or children. You only have seen the light reflected from them. You only see light passing through your eyes' lenses, stimulating the rods and cones in your retina, and transmitting the forms and colors of those you love to your brain. Just as you enjoy seeing your loved ones from the light they reflect, you can find joy seeing divine light reflected from every place you look. This set of instructions teaches you how to see the spectrum of divine light through your camera lens.

PHOTOGRAPH GOD IN REFLECTIONS OF REALITY

God does not exist in reality. God is reality itself. Rabbi David Aaron, who teaches *kabbalah* in the Old City of Jerusalem, explains in his book *Seeing God*[33] that God is the all-embracing context for everything. In Hebrew, God is called *Hamakom*, which means "The Place." God is the place where everything is happening. You do not exist alongside God. You exist within God, within the only one reality that is God. Everything is in God, God is in everything, but God is also beyond everything.

Seeing God is all about getting in touch with reality. If you want to photograph God, focus your lens on *Hamakom*, The Place, anyplace where you see divine light illuminating reality. Let your camera collect the light reflecting from the reality shaping your everyday life and you will find yourself photographing God in action.

To photograph God as the place where all action takes place, you need to redefine the English word "God," a Germanic word conjuring up images of some all-powerful being zapping us if we step out of line. This is an alien concept in *kabbalah*. Names for God in biblical Hebrew are not really names for God at all. They are names linked to divine attributes, to each of the colors of the spectrum of divine light, to divine roles rather than essence. Hebrew speakers call God *Hashem*, literally "The Name." When you see the word "God" here, read it as the Hebrew word *Hashem*, or even *Habili Shem* "The Nameless One" forming all of reality and beyond.

PHOTOGRAPH GOD AS A VERB

God is a verb. God is no thing – *no*thing in the process of becoming everything. The great 16th-century kabbalist, Rabbi Isaac Luria, known as The Ari, calls God *Ha'efes Hamukhlat*, "The Absolute Nothingness," as well as *Ein Sof*, "Endless." God is One, infinite nothingness and everything in the universe all at once, absolutely invisible becoming visible wherever we focus our lens.

You can discern God over time, in the flow, in the action, in the process of something becoming something else. The primary biblical divine name *YHVH*, usually translated as "God," should really be translated as "Is-Was-Will Be." *YHVH* integrates past, present, and future of the verb "to be." It is associated with the divine attribute of inner beauty (*tiferet*). When beauty hidden in the mundane suddenly jumps out at you, catch the action in a series of photographs of Is-Was-Will Be. Don't freeze the action in a single still-life picture, *nature morte* ("dead life" in French). Photograph living processes in a series of images like comic strip or storyboard sequences. Show before and after.

PHOTOGRAPH *KUZU*

KUZU is *YHVH* in motion. The biblical passage beginning with "Hear, O Israel, *YHVH* is our God, *YHVH* is One" (*Deuteronomy 6:4–9*), is written by a scribe on small parchment scrolls affixed to doorposts in Jewish homes. These mini-Torahs called *mezuzot*, a word derived from the root *zaz*, which means to move. Each scroll is rolled up with the biblical text on the inside. On the outside of the scroll at the place on the reverse side of where *YHVH* is written, the scribe writes *KUZU* to set God in motion. *K-U-Z-U* is spelled with each of the four letters that follow *Y-H-V-H* in the Hebrew alphabet. *K* follows *Y*, *U* follows *H*, *Z* follows *V*, and *U* follows *H*. It is as if we were to write GOD as HPE, H being the letter following G, P the letter following O, and E the letter following D. In addition to moving each of the letters in *YHVH* forward, *KUZU* is written upside-down to invite us to see God in motion from multiple viewpoints. Photograph *KUZU*. God becomes even more active in the kabbalist's prayer book. *KUZU* follows *TDHD*, the four Hebrew letters preceding *YHVH*, as if GOD moves backwards to FNC and forwards to HPE. Photograph *TDHD* and *KUZU*, before and after. Photograph God dancing back and forth.

PHOTOGRAPH GOD IN EVERY NOOK AND CRANNY OF LIFE

Look for God in every nook and cranny of your life. Rabbi Joseph B. Soloveitchik,[34] one of the foremost thinkers of the 20th century, teaches that you should not direct your glance upward but downward, not aspire to a heavenly transcendence nor seek to soar upon the wings of some abstract, mysterious spirituality, but to fix your gaze upon concrete reality. He emphasizes that you should not confine your search for God to houses of worship for you can find God penetrating into every nook and cranny of life. Photograph God in the details of empirical reality permeating your daily mundane activities.

PHOTOGRAPH GOD IN YOUR WORK AND SOCIAL LIFE

Photograph God into everything you do. The Lubavicher Rebbe, Menachem M. Schneerson,[35] the greatest contemporary Hasidic master, emphasizes that it is not enough to rest content with your own spiritual ascent, the elevation of your soul in closeness to

God. You must also strive to draw spirituality down into the world and into every part of your involvement with it – your work and your social life – until not only do they not distract you from your pursuit of God, but they become a full part of it. Photograph God in your work and in your relationships with others.

PHOTOGRAPH GOD AT GROUND LEVEL
In his acclaimed novel, *The City of God*,[36] E. L. Doctorow echoes these rabbinic thoughts: "If there is a religious agency in our lives, it has to appear in the manner of our times. Not from on high, but a revelation that hides itself in our culture, it will be ground-level, on the street, it'll be coming down the avenue in the traffic, hard to tell apart from anything else. It will be cryptic, discerned over time, piecemeal, to be communally understood at the end like a law of science." Photograph God everywhere you go and in everything you do.

PHOTOGRAPH GOD IN THE STILL SILENT VOICE
Living for seven years in the Negev desert, I would frequently walk through the desert mountains where a strong silence surrounded me and followed me. The silence grew more intense as I stopped to stoop down to get a close look at a tiny flowering plant emerging from the crevice of a rock. The Hebrew word for "desert" *MiDBaR* is spelled with the same letters as the word for "speaking" *MiDaBeR*. The desert speaks softly about delicate forms of life. In the desert, you can see the quiet voice of God. In the Sinai desert, "all the people saw the sounds" (*Exodus 20:15*) rather than heard them.

Standing on a desert mountain, the prophet Elijah saw a great powerful wind, smashing mountains and breaking rocks. After the wind came an earthquake and after the earthquake was fire and after the fire there was a still silent voice. Elijah saw God in the still silent voice, rather than in the mighty wind, rather than in the rumbling earthquake, rather than in the raging fire (*I Kings 19:11–12*). Listen for the still silent voice as you photograph God in the intimate spaces and minute details of your life. Transform your vision of small ordinary events into extraordinary images.

PHOTOGRAPH THE SPECTRUM OF DIVINE LIGHT
Just as a prism breaks up white light into the colors of the spectrum, *kabbalah* reveals a spectrum of divine light derived from the biblical passage "You God are the compassion, the strength, the beauty, the success, the splendor, and everything in heaven and on earth" (*Chronicles 1:29*). Look for these six attributes of divine light flowing down into your life.

Hesed: Compassion / Largess / Loving All
Gevurah: Strength / Judgment / Setting Limits
Tifert: Beauty / Aesthetic Balance / Inner Elegance

Netzakh: Success / Orchestration / Eternity
Hod: Splendor / Gracefulness / Magnificence
Yesod: Foundation / Everything Integrated / Gateway to Action

Focus on episodes expressing these attributes as you walk through the streets, ride on a bus, shop in the mall, dance at a wedding, hike in the countryside, or come home from work. "God walks in the midst of your camp" (*Deuteronomy 23:15*). *KeReV*, the Hebrew word for "midst," shares the same root as being "close," *KaRoV.* As you sense the closeness of God walking with you, create six sets of pictures revealing the spectrum of divine light that you see all around you.

LET GOD LOOK BACK AT YOU
Photographer Jan Phillips quotes from Rabbi Elimelech as she shares her thoughts about focusing her lens on God in her book on photography and creativity, *God is at Eye Level*: "Whoever does not see God in every place does not see God in any place[…] My eyes find God everywhere, in every living thing, creature, person, in every act of kindness, act of nature, act of grace. Everywhere I look, there God is looking back, looking straight back."[37]

My students' work that I posted on the *Photograph God* blog includes Karen's photographic sequence expressing *compassion* as a process that begins with hungry feral cats, hungry for love and food, surrounding an elderly gentleman who has seen much in his life who chose to respond to their hunger. He pets them in one photograph, satisfying their hunger for love, and in the next photograph portions out food for each of them making sure there is enough for all. Sharon sees *compassion* as the divine loving kindness bestowed upon a bride on her wedding day. Her photographs show a beautiful bride, her eyes closed in contemplation, enveloped in the aura of her new husband's love, as they stand close together under a wedding canopy.

Dalia sees *success* as the *victory* of good over evil and the love of the Jewish people for its Torah for *eternity*. As a participant, she photographed the "March of the Living" as she and her fellow students from Israel walk together to Nazi death camps in Poland in order to never forget the horrible nightmare and unimaginable suffering of millions of Jews brutally murdered there. On her return home to Israel, she photographed *strength* as her brave peers, soldiers defending their country against its current enemies seeking to destroy it. They are wrapped in prayer shawls reading from a Torah scroll in an open field marking the beginning of their dangerous day.

Esti documents avian *strength* in a photographic sequence showing a parrot chick pecking its way out of its egg and avian *splendor* as the metamorphosis of the young parrot, a strange-looking earthbound creature with stubby feathers, into a magnificent bird in flight. Roni's photographic sequence shows the birthing of a calf at a dairy farm on Israel's coastal plain, an awesome event expressing *beauty* as helping to bring new life into the world. It reveals *beauty* as the vital balance between the farmer's *compassion* aiding a cow in labor, and the *strength* of his arms pulling the calf through the birth canal.

The biblical prophet Zechariah envisioned a beautiful future during the depths of despair when Jerusalem was razed by its enemies and the Jewish people exiled. Tzipi sees Zechariah's vision being realized in our day after two millennia of bitter exile. "Thus said God: I will bring My people from the land of the east and from the land where the sun sets to dwell within Jerusalem [...] We will see the wondrous vision of elderly men and women once again sitting in the streets of Jerusalem and the streets of the city will be filled with boys and girls playing." (*Zechariah 8:6–7*). Tzipi photographed *beauty* as she and her brother sitting with their great-grandparents, both 91 years old, in their home in Jerusalem. Her grandparents have 7 children, 47 grandchildren, 170 great-grandchildren, and 6 great-great-grandchildren – in total 230 offspring!

Notes

1. Don Tapscott and Anthony D. Williams, *Wikonomics: How Mass Collaboration Changes Everything* (New York and London: Portfolio, 2008), p. 15, 3.
2. Menache M. Schneerson, *Toward a Meaningful Life: The Wisdom of the Rebbe*, adapted by Simon Jacobson (New York: William Morrow, 1995), p. 191.
3. Douglas Rushkoff, *Nothing Sacred: The Truth about Judaism* (New York: Three Rivers Press, 2004).
4. Douglas Rushkoff, *Media Virus* (New York: Ballantine Books, 1996).
5. Douglas Rushkoff, *Cyberia: Life in the Trenches of Hyperspace* (San Francisco: HarperSanFrancisco, 1994).
6. Public Broadcasting Company, *Digital Nation: Life on the Virtual Frontier* (PBS, 2010), http://www.pbs.org/wgbh/pages/frontline/digitalnation/us/. Accessed October 10, 2010.
7. Martin Buber, *Between Man and Man*, trans. Ronald Gregor Smith (London and Glasgow: Collins Fontana Library, 1969), p. 146.
8. *Wikipedia*, "Tribe", http://en.wikipedia.org/wiki/Tribe. Accessed October 10, 2010.
9. Joel Kotkin, *Tribes: How Race, Religion and Identity Determines Success in the New Global Economy* (New York: Random House, 1994).
10. Seth Godin, *Tribes: We Need to Lead You* (New York: Portfolio/Penguin, 2008).
11. Bo Leuf and Ward Cunningham, *The Wiki Way: Quick Collaboration on the Web* (Indianapolis: Addison-Wesley, 2001).
12. Don Tapscott and Anthony D. Williams, *Wikinomics: How Mass Collaboration Changes Everything* (New York: Portfolio/Penguin, 2008).
13. Thomas H. Carefoot, "Toward the New (Isopod) Millennium," *Israel Journal of Zoology*, 44: 3–4, 1998.
14. Max Hecht (ed.), *Evolutionary Biology*, 33 volume series (Luxembourg: Springer, 1967–2000).
15. Mel Alexenberg, "Pigmy 'Armadillos': The Pill-bugs," *Junior Natural History Magazine*, vol. 26, no. 10, February 1962.
16. Mel Alexenberg and Miriam Benjamin, "Legacy Thrones: Intergenerational Collaboration in Creating Multicultural Public Art," in Angela M. La Porte (ed.), *Community Connections:*

Intergenerational Links in Art Education (Reston, VA: National Art Education Association, 2004); and Mel Alexenberg and Miriam Benjamin, "Creating Public Art Through Intergenerational Collaboration," *Art Education*, 57:5, September 2004.
17. Suzi Gablik, *The Reenchantment of Art* (New York: Thames and Hudson, 1991), p. 5.
18. New World School of the Arts is a joint venture of the University of Florida, Miami Dade College, and Miami Dade Public Schools. Mel Alexenberg served as Dean of Visual Arts at NWSA from 1990–2000. He created there the first BFA program in Environmental Public Art in the USA.
19. Charles Green, *The Third Hand: Collaboration in Art from Conceptualism to Postmodernism* (Minneapolis: University of Minnesota Press 2001), p. x.
20. Susan Perlstein, S. Charnow and E. Nash, *Life Review Training Manual* (Brooklyn, NY: Elders Share the Arts, 1994).
21. Eric H. Erikson, J. M. Erikson and H. Q. Kivnik, *Vital Involvement in Old Age* (New York and London: Norton 1986).
22. R. L. Jones, Jr., "Modern and Postmodern: Questioning Contemporary Pedagogy in the Visual Arts," in J. Hutchins and M. Suggs (eds.), *Art Education: Content and Practices in a Postmodern Era* (Reston, VA: National Art Education Association, 1997), pp. 91–102.
23. Tom Anderson, "Toward a Postmodern Approach to Art Education," in J. Hutchins and M. Suggs (eds.), *Art Education: Content and Practices in a Postmodern Era* (Reston, VA: National Art Education Association, 1997), pp. 62–73.
24. A. Efland, F. Freedman and P. Stuhr, *Postmodern Art Education: An Approach to Curriculum* (Reston, VA: National Art Education Association, 1996).
25. Ibid., p. 72.
26. H. Pearse, "Doing Otherwise: Art Education Praxis in a Postparadigmatic World," in J. Hutchins and M. Suggs (eds.), *Art Education: Content and Practices in a Postmodern Era* (Reston, VA: National Art Education Association, 1997), pp. 31–39.
27. Lucy R. Lippard, *Mixed Blessings: New Art in a Multicultural America* (New York: Pantheon, 1990), p. 5.
28. Lawrence Halprin and Jim Burns, *Taking Part: A Workshop Approach to Collective Creativity* (Cambridge, MA: The MIT Press, 1974), p. 11.
29. Lippard, *Mixed Blessings*, p. 4.
30. Boris Groys, *Art Power* (Cambridge, MA: The MIT Press, 2008).
31. Stefan Sonvilla-Weiss, *(In)Visible: Learning to Act in the Metaverse* (New York: SpringerWien, 2008).
32. Mel Alexenberg, "Art with Computers: The Human Spirit and the Electronic Revolution," *The Visual Computer: International Journal of Computer Graphics*, 4: 1, 1988, pp. 1–3.
33. David Aaron, *Seeing God: Ten Life Changing Lessons of the Kabbalah* (New York: Berkley Books, 2001).
34. Joseph B. Soloveitchik, *Halakhic Man*, trans. Lawrence Kaplan (Philadelphia: Jewish Publication Society of America, 1983), p. 99.
35. Menachem M. Schneerson, *Torah Studies*, adapted by Jonathan Sacks (London: Lubavitch Foundation, 1986), p. 320.
36. E. L. Doctorow, *City of God* (New York: Plume Book/Penguin Putnam, 2001), p. 254.
37. Jan Phillips, *God is at Eye Level: Photography as a Healing Art* (Wheaton, IL and Chennai, India: Quest Books, 2000), p. 141.

Chapter 6

Halakhic Perspectives: *Creating a Beautiful Life*

Halakhic man is a man who longs to create, to bring into being something new, something original. The dream of creation is the central idea in halakhic consciousness – the idea of the importance of man as a partner of the Almighty in the act of creation, man as creator of worlds. This longing for creation and the renewal of the cosmos is embodied in all of Judaism's goals.[1]

Joseph B. Soloveitchik

It is not enough for the Jew to rest content with his own spiritual ascent, the elevation of his soul in closeness to G-d. He must also strive to draw spirituality down into the world and into every part of his involvement with it – his work and his social life – until not only do they not distract him from his pursuit of G-d, but they become a full part of it.[2]

Menachem M. Schneerson

Halakhic perspective invites us to use our creativity in drawing spirituality down into our gross material world in ways that bring beauty and goodness into our lives and the lives of others. Although both halakhic and kabbalistic perspectives place the creative process in an honored position, halakhic perspective only values human creativity that acts to couple beauty with goodness. Good and beautiful, *tov v'yafeh*, are an inseparable pair of words linked to the Jew's daily affirmation of divine unity in the morning service. Halakhic perspective sees the stage for integrating beauty and goodness, not in some transcendental realm, but in every aspect of our concrete existence. It seeks to transform spiritual values in an ideal world into physical forms that draw beauty and goodness down into everyday life. Certainly, the arts provide an especially fertile arena for realizing this process of concretizing spiritual values.

Halakhic perspective originates in the biblical verse that begins the story of the Hebrew nation. "God said to Abram, 'Walk yourself away from your land, from your birthplace, from your father's house to the land that I will show you'" (*Genesis 12:1*). The Hebrew word for "walk" is the root of the word *halkhah*, a walkway to a beautiful life. Tradition tells us to read *halakhah* as *halikhah*, "walking." The name of God addressing Abram, *YHVH* (Was-Is-Will Be), is associated with the core of the creative process, the *sepherah* of Beauty. God tells him that he can freely come to see a land of beauty by walking away from the all too familiar ways of the idolatrous world into which he was born. Halakhic perspective follows a map for avoiding dead-end paths and reaching ways that open up opportunities for creating a beautiful life. "Its ways are ways of pleasantness and all its paths are peace" (*Proverbs 3:17*).

The relationship between Torah, the divine revelation, and its creative development through *halakhah* is analogous to a fixed map of divine creations – continents, oceans, rivers, mountains and valleys – on which human beings, God's human partners in creation, draw countries, cities, highways and streets. The divine revelation began in God's relationships with Abraham, Isaac and Jacob, reached a high point with the revelation at Mount Sinai, was clarified by the Prophets, and has been creatively interpreted ever since. This creative dynamic is documented in the Talmud and in ongoing rabbinic dialogues spanning past, present and future.

There is a revealing story told in the Talmud about a group of rabbis debating whether an oven that had become impure could be purified. They all agreed it could not be purified with the exception of the great scholar Rabbi Eliezer. Rabbi Eliezer tried in every rational way to convince his colleagues that his opinion is correct but to no avail. All the other sages disagreed with him. In utter frustration at being certain he was right, he said to them, "If the *halakah* is according to me, let that carob tree prove it." He pointed to a nearby tree that moved from its place and replanted itself a distance away. The sages did not see any connection between the moving tree and his arguments. "A moving tree proves nothing," they said to him. Rabbi Eliezer then tried to prove his point by making a stream of water reverse direction. The sages remained unconvinced. Rabbi Eliezer in desperation called out to God to prove him correct. A voice from Heaven responded saying that Rabbi Eliezer's way is the right way. Then Rabbi Joshua spoke up, "It is not in the Heavens" (*Deuteronomy 30:12*). Rabbi Jeremiah affirmed Rabbi Joshua's refusal to heed God's words, "Since the Torah has already been given to us at Mount Sinai, we do not pay attention to a Heavenly voice for You have written in Your Torah, 'Decide according to the majority' (*Exodus 23:2*)." When Rabbi Nathan met the prophet Elijah, he asked him what was the Holy One, Blessed be He, doing when all this was going on. Elijah answered, "He was laughing and saying, 'My children have defeated me, my children have defeated me.'"[3]

This insightful story highlights the halakhic perspective as a creative encounter with the divine gift of Torah. God laughed as a good parent does when his children grow up to become independent. Rabbi Soloveitchik explains, "Halakhic man received the Torah from Sinai, not as a simple recipient but as a creator of worlds, as partner with the Almighty in the act of creation. The power of creative interpretation (*hiddush*) is the very foundation of the received tradition."[4]

Beyond religion and science

Rabbi Soloveitchik compares halakhic man to two other types – *homo religiosus* and cognitive man – admitting the difficulty of analyzing halakhic man's religious consciousness by applying methodologies of philosophy and psychology of religion. Halakhic man is a type that is unfamiliar to students of religion. He is more like the cognitive man of science and mathematics than the spiritual man of religion who seeks God in heaven. *Homo religiosus* and halakhic man travel in opposite directions. *Homo religiosus* starts out in this imperfect world and ends up in supernal realms seeking an uncontaminated heavenly existence. In contrast, halakhic man starts out in supernal realms and ends up in this gross material world where he acts to transform it into a better place by bringing transcendence down into every aspect of life.

> Halakhic man's relationship to transcendence differs from that of the universal *homo religiosus*. Halakhic man does not long for a transcendent world, for "supernal" levels of a pure, pristine existence, for was not the ideal world – halakhic man's deepest desire, his darling child – created only for the purpose of being actualized in our real world? It is this world which constitutes the stage for halakhah, the setting for halakhic man's life. It is here that the halakhah can be implemented to a greater or lesser degree. It is here that it can pass from potentiality into actuality. It is here, in this world, the halakhic man acquires eternal life! "Better is one hour of Torah and *mitzvot* in this world than the whole life of the world to come," stated the tanna in Avot [4:17], and this declaration is the watchword of the halakhist.[5]

When *homo religiosus* gazes at the myriads of objects and events in the cosmos, he has no interest in discovering the secrets of creation and expressing them in sets of mathematical equations. He is intrigued by the mystery of existence and does not want to dispel it. The obscure and concealed in nature stimulate spiritual feelings of awe that would be lost if they were clearly understood. The transcendent approach to reality that attempts to find in the material world traces of higher worlds is the primary trait of the man of God. Since *homo religiosus* sees this world as a pale image of another world, he constantly searches for an existence above empirical reality. His unhappiness and dissatisfaction with this world pushes him to seek a better place in some transcendental realm.

> The ethical and religious ideal of *homo religiosus* is the extrication of his existence from the bonds of this world, from the iron chains of empirical reality, its laws and judgments, and its elevation up to the level of being a higher man, in a world that is wholly good and wholly eternal.[6]

Cognitive man, on the other hand, delimits his quest to understanding empirical reality through scientific methods that reveal the laws of nature. His worldview is contained within the limits of concrete physicality that invites scientific understanding. Cognitive man pays no attention and has no interest in worlds that may exist above the rule of empirical reality or totally denies that such worlds exist. He studies the material world in order to create conceptual models that exhibit congruence between his model and empirical reality.

Although halakhic man differs in his worldview from both *homo religiosus* and cognitive man, he more closely resembles cognitive man, a combination of mathematician and scientist. The mathematician creates an ideal world that invites scientists to establish relationships between his ideal world and the world of concrete reality. The mathematician constructs an ideal, lawful unified system that does not require any validation of its truth and validity except from within mathematics itself. Mathematics does not have to address concrete phenomena. Scientists use mathematical models to seek parallelisms between them and empirical reality. Like the ideal constructs in mathematics, *halakhah* begins with a divinely sanctioned ideal world. Halakhic man is on the one hand like the mathematician whose creative play leads to new relationships within the realm of mathematics. He studies Talmud to discover fresh creative understandings and new interrelationships within the halakhic discourse itself. On the other hand, halakhic man is like the scientist who looks outside at his physical environment to uncover relationships, parallelisms, and congruencies between an ideal construct and concrete reality. Employing human creativity to reveal innovative ways to relate ideal halakhic constructs to life as it is lived in the material world is the highest ideal of Judaism.

There is no object or phenomenon in this material world that the *a priori halakhah* does not approach with its ideal standard. Transformation of the profane world into holy acts is dependent upon natural cosmic phenomenon, the first light of dawn, the setting sun, the appearance of the first sliver of a new moon, the first stars dotting the night sky, a spring bubbling quietly, animals' different ways of eating, trees issuing their first fruits, and the size of an olive. *Halakhah* encompasses state, government, relationships of individuals in a communal context, war and peace, business partnerships, legal rights, and every other aspect of society. *Halakhah* clarifies and elucidates aspects of family life – marriage, divorce, sexual relations, and the rights, obligations, and duties of husband and wife.

Since *halakhah* relates to the whole of reality in all of its fine and detailed particulars, it also addresses art. Rabbi David Avraham Spector compiled *halakhic resposa*, questions about art addressed to rabbis through the ages and their responses, in his book *Shut Omanut* (Respona on Art).[7] There are sections on painting and sculpture of human beings, landscapes, dragons, astronomical bodies and animals. Other sections deal with nudity and modesty in art, Rubens' paintings, visiting art galleries, encountering sculptures of idols in museums, making paintings of images of important biblical and rabbinic figures, studying art history, and using symbols on ritual objects. Sixty pages are devoted to computer graphics from the digital manipulation of scriptural passages and divine names to using computers on the intermediate days of holidays.

Halakhah not only addresses art directly, but it explores related experiences that elucidate issues in postmodern art. One of the major issues of postmodern thought is how meaning is created through context. In the chapter on semiotic perspective, I related the Talmudic discussion about how a hand and foot from a broken idolatrous statue can be either permitted or prohibited according to context. It parallels contemporary discussions about how Brillo boxes were transformed by Andy Warhol in his artwork *Brillo Boxes* by stacking them in an art gallery rather than in a supermarket. Context determines whether the boxes are artworks or mere real things. An idolater smashing an idolatrous statue nullifies it as an idol thereby rendering the hand and foot as being permitted for a home decoration or plaything for children. However, if a Jew smashes the same statue, the same hand and foot is prohibited because one who did not consider it as being holy in the first place could not nullify its idolatrous character. If the statue topples by itself, the hand and foot are permitted because the owner of the idol annuls it by saying, "If it could not save itself, so how could it save me?" The hand and foot mounted on a pedestal show that they are still being honored and are, therefore, prohibited.

The major halakhic discussions related to the art experience explore how to embellish all of our deeds in beautiful ways, an aesthetic process called *hidur mitzvah*. The sublime beauty that *homo religiosus* seeks in transcendent realms and the elegant forms and relationships in nature that cognitive man reveals through scientific research are applied to every aspect of day to day living by halakhic man. *Halakhah* is a creative process to transform mundane experiences into miraculous events. Through a practice called *tikun olam*, it reshapes perceptions of all that is common and ordinary in our lives into extraordinary visions of profound beauty flowing from our creative attempts to embed compassion, splendor, and grace into our gross material world.

Lesson from 9/11: choose life not death

Halakhic perspective recognizes that same creative process that makes human beings partners of God in *tikun olam*, making the world a better place, can be hijacked for evil purposes. It recognizes that creative energies can be perverted into seeing evil acts as being beautiful and into perpetrating malevolent behavior shrouded in aesthetics. It also recognizes that those who are apathetic to the evil acts of others are in the same camp as those who perpetrate evil. Halakhic perspective looks askance at those who seek ideal beauty in heavenly realms and in theories of aesthetic purity while doing nothing practical to alleviate ugliness in our concrete environment. Biblical commentator Nehama Leibowitz points out that the admonition in *Leviticus 19:14* not to put a stumbling block before the blind has been understood by the sages of the Jewish people as an imagery concept generalized to refer to all cases of moral turpitude resulting from apathy to evil acts.

> The Torah teaches us that even by sitting at home doing nothing, by complete passivity and divorcement from society, one cannot shake off responsibility for what is transpiring in the world at large, for iniquity, violence and evil there. By not protesting, you have become responsible for any harm arising therefrom, and have violated the prohibition: "Thou shalt not put a stumbling block before the blind."[8]

We are charged with actively integrating the good and the beautiful through life-affirming actions. The Torah gives us free will: "Today I have set before you a choice between life and good and death and evil [...] Choose life!" (*Deuteronomy 30:15, 19*). The Talmud teaches:

"The saving of a life overrides the commandments of the entire Torah. He should live by them and not die by them."[9]

Jews have always been the prime enemies of death-loving people from the ancient Egyptian cult of the dead to fascist Europe and the suicide bombers of militant Islam. Art critic Walter Benjamin wrote about the anti-Semitic culture of Europe: "Its self-alienation has reached such a degree that it can experience its own destruction as an aesthetic pleasure of the first order." He wrote these words in 1935 as he fled the Nazi terror. They appear in his prophetic essay that is informing much of postmodern thought, "The Work of Art in the Age of Mechanical Reproduction." Benjamin shows how industrial age technology inspired Italian Futurism to strive toward creating an art of movement that expressed death and violence in aesthetic terms by quoting from the manifesto of the fascist Futurist artist Marinetti:

> War is beautiful because it establishes man's dominion over subjugated machinery by means of gas masks, terrifying megaphones, flame throwers, and small tanks. War is beautiful because it initiates the dream of metalization of the human body. War is beautiful because it enriches a flowering meadow with fiery orchids of machine guns. War is beautiful because it combines the gunfire, the cannonades, the cease-fire, the scents, and the stench of putrefaction into a symphony. War is beautiful because it creates new architecture, like that of big tanks, the geometrical formation of flights, the smoke spirals from burning villages [...] Poets and artists of Futurism! Remember these principles of an aesthetics of war so that your struggle for a new literature and a new graphic art may be illuminated by them![10]

The toppling of the Twin Towers presented a worldwide audience with the 21st century's most powerful media event. Millions watched real-time TV images of a jumbo jet crashing into one tower while people were leaping to their deaths from the other flaming tower to escape the inferno. The Saudi Arabian and Egyptian hijackers enthusiastically gave their lives to create a magnificently potent media image that would be etched in the minds of humanity for all time. Their media event was a commanding semiotic statement attesting to the strength of militant Islam in toppling the symbols of haughty America's economic and military power – the World Trade Center and the Pentagon. I watched Palestinian Authority television broadcasting to martial music mobs of frenzied revelers dancing in the streets firing their assault rifles in the air in response to ongoing replays of the wonderfully horrific TV images of death and destruction. One ecstatic young Arab who was interviewed reiterated that the Twin Towers in flames was the most beautiful sight that he had ever seen. An enterprising artist created a sculpture of the Twin Towers in flames that was mass-produced and quickly became the best selling toy in Gaza City along with dolls of their hero Osama Bin Laden.

Simultaneously, an art exhibition was mounted at An-Najah University in Nablus. Visitors pushed to see realistically sculpted body parts and pizza slices strewn throughout an environment set for a performance artwork. The artists had constructed a replica of the Sbarro Pizzeria in the heart of Jerusalem in which men, women, children, and entire families were mercilessly slaughtered by an Arab suicide bomber. Wearing a terrorist's military uniform and black mask, the performance artist entered the mock pizzeria under a sign "Kosher Sbarro" and set off a simulated explosion to the cheers of the crowd. Upon entering and leaving, the visitors enthusiastically wiped their feet on Israeli and American flags used as doormats. They then encountered a mannequin outfitted as a terrorist standing next to a large boulder. A recording placed behind the rock called out in Arabic: "O believer, there is a Jewish man behind me, come and kill him." In another room, two students dressed as suicide bombers, each with the Koran in one hand a Kalashnikov assault rifle in the other hand, were reenacting the grisly last testaments in front of a video camera that suicide bombers create before carrying out their deadly attacks. The real-life last testaments have become a video folkart broadcast on Palestinian Authority television as homage to the killers. The videos are being shown proudly by the killers' families who express their happiness that their sons are enjoying an afterlife surrounded by harems of virgins in reward for their beautiful act of killing themselves in order to kill Jews.

As in biblical times, the Land of Israel is once more plagued with the disciples of Molekh and Baal who spill innocent blood and burn their children alive. Biblical consciousness abhors such abominable behavior (*Leviticus 18:21, 20:2–5*). The prophet Jeremiah laments that there are those who "have built high platforms in the Valley of Hinnom to burn their sons and daughters in fire' (*Jeremiah 17:31*). The Valley of Hinnom, which Jeremiah calls the "Valley of Slaughter," lies midway between Gaza City and Nablus, about an hour's drive from these two new centers of Molekh worship. "They filled this place with blood of innocent people; they built the high platforms of the Baal to burn their sons in fire as burnt-offerings" (*Jeremiah 19:4–5*). The Hebrew for the "Valley of Hinnon" is *gehinnom*, used metaphorically as the appellation for hell.

The 20th century's foremost expert on halakhic perspective, Rabbi Joseph B. Soloveitchik writes in his seminal book *Halakhic Man*:

> Many religions view the phenomenon of death as a positive spectacle, inasmuch as it highlights and sensitizes the religious consciousness and "sensibility." They, therefore, sanctify death and the grave because it is here that we find ourselves at the threshold of transcendence, at the portal of the world to come. Death is seen as a window filled with light, open to an exalted, supernal realm. Judaism, however, proclaims that coming in contact with the dead precipitates defilement. Judaism abhors death, organic decay, and dissolution. It bids one to choose life and sanctify it. Authentic Judaism as reflected in halakhic thought sees in death a terrifying contradiction to the whole of religious life. Death negates the entire magnificent experience of halakhic man.[11]

It is customary Jewish practice to bury a dead man in his *talit* prayer shawl with the *tzitzit* fringes removed. Removing the *tzitzit* symbolizes that he can no longer act out any of the precepts of the Torah, that he has lost the opportunity to create and make the world a better place, and that he is powerless to change anything at all. Death robs a Jew of every opportunity to draw spirituality down into the material world. Spirituality cannot exist without a physical body for its realization.

The Talmud tells us that when Moses ascended on high to receive the Torah, the ministering angels spoke to the Holy One questioning how such a secret treasure could be given to lowly creatures of flesh and blood instead of to them. God deferred to Moses to respond to the angels. Moses answered,

> In the Torah it is written, "I am *YHVH*, your God, who brought you out of the land of Egypt." Did you go down to Egypt? Were you enslaved to Pharaoh? It is written, "Remember the Sabbath day to keep it holy." Do you work hard all week that you need to rest? It is written, "Honor your father and mother." Do you have fathers and mothers? Again it is written in the Torah, "Thou shalt not murder. Thou shalt not commit adultery. Thou shalt not steal." Is there jealousy among you? Are there evil impulses among you?

The angels immediately conceded.[12] The *midrash*[13] about Abraham's decision to make a barbeque for his wife and guests rather than succumbing to the delusion of finding a readymade Garden of Eden teaches that we ourselves have the power to create heaven or hell at home with our spouse and children and in our relationships with friends, neighbors and strangers. A vision of Paradise far off at the end of a cave or in some heavenly realm above is either a mirage or a fraud. We enter the Garden of Eden through the compassionate quality of our everyday encounters with others. Halakhic perspective charges us with creating a beautiful life in the here and now.

Tower of Babel: disastrous creativity

According to the Bible, the first creative architectural collaboration of humanity was disastrous. "Come, let us build a city and a tower with its top in the heavens and let us make a name for ourselves, lest we be dispersed across the whole earth." (*Genesis 12:4*) The builders of the Tower of Babel mistakenly thought they could work together to find God by ascending to meet Him in heaven. The next sentence in the Bible begins, "God descended to look at the city and tower which the sons of man built" (Genesis 12:5). Rabbi Kook[14] points out that we need to open our eyes to see divinity descending into our immediate surroundings rather than to search for God in some far off heavenly realms. Holiness and sanctity exists here in the physical world. The narrative of the Tower of Babel is immediately followed by the story of Abraham who was the first to mend the rift between heaven and earth by showing his generation ways to fuse the physical and the spiritual here in this world.

In an arrogant expression of power over nature, the builders made bricks of clay and fired them to make them hard as rock. They used these man-made bricks rather than stone, the common building material of the times supplied by nature. They valued the prized production of their hands above all else. The *midrash* presents a narrative to explain the basis of divine displeasure at the moral decay that results from fervently focusing on the material world divorced from its spiritual elements and from exclusively focusing on the means at the expense of the end.

> Many years were spent building the Tower. It reached so great a height that it took a year to mount to the top. A brick was, therefore, more precious in the sight of the builders than a human being. If a man fell and died they paid no attention to him; but if a brick fell down they wept because it would take a year to replace it.[15]

The offense of "let us make a name for ourselves" is added to the offenses of attempting to find the spiritual in heaven rather than here on earth and of valuing the work of human hands above human life. If all humanity that survived the Flood acted together in the Tower-building project for the purpose of self-aggrandizement, to whom is "let us make a name for ourselves" addressed? It could not be directed to some other group of people, another community or nation, since none existed at that time. It is directed to the individual and to God. It diminishes the individual by elevating the collective above him. "Us" and "ourselves" refers to the community of builders who see the might of the collective against the individual who is subordinate to the group or against God who hides high above threatening a new Flood. The builders were apathetic when one of them fell to his death. They shot arrows from the top of the Tower straight up into the sky hoping to find God's blood on their arrows as they fell back to earth. The individual was but a dispensable cog on the Tower-building machine. God was a threat that they aimed at bringing down to earth dead so they would reign in heaven from the top of their Tower. The Tower of Babel story aims at developing biblical consciousness that values community when its purpose is not self-aggrandizement, but aggrandizement of God and of each individual created in the Divine image. A community's worth is determined by how successful it is in honoring and serving the individual and how successful it is in bringing God down to earth alive.

The greatest transgression of the Tower builders, however, was their defying the Divine will that expects humanity to revere and applaud differences between peoples. In postmodern terms, they failed to realize and honor the values of polyculturalism. It is most significant that the Bible, which does not waste words, repeats the same message three times, each time in relation to one of the descendents of Noah's three sons. After naming the 14 nations emerging from Japheth, we read: "From these the islands of the nations were separated in their lands – each according to its language, by their families, in their nations" (*Genesis 10:5*). After naming the 30 nations from Ham, "These are the descendants of Ham, by their families, by their languages, in their lands, in their nations" (*Genesis 10:20*). And after naming the 26 nations from Shem, "These are the descendents of Shem according to their families, by their languages, in their lands, by their nations. These are the families of Noah's descendants, according to their generations, by their nations; and these nations were separated and spread across the earth after the Flood" (*Genesis 10:31–32*). Rabbi Samson Raphael Hirsch[16] comments that it is not without good reason that God makes people so different from one another. Individuals are intended to compliment one another. Just as God did not create a single mold in which to cast identical clones, so each of the biblical seventy nations of the world was not meant to come together to speak one language, to share a common set of cultural values, and to engage in a singular mission of self-aggrandizement. "God descended to see the city and the tower that the son of man had built... From that place, God scattered them all over the face of the earth, and they stopped building the city" (*Genesis 11:5, 8*). Each nation has its unique and distinct voice to contribute to the grand planetary choir singing God's praises.

Dialogic art of reconstructive postmodernism honors collaboration and collective creativity, but not of the Tower-builders kind. It facilitates expressions of diversity within participatory group processes. As in the collective creativity workshops run by Lawrence Halprin, they allow group differences to *emerge*, not *submerge*.[17] Although it is a global enterprise, the postdigital wikiworld of collaboration avoids the disastrous collective creativity of Babel by welcoming the originality and initiative of the participants. The thousands of volunteers who have been creating *Wikipedia*, for example, have not created a single-language authoritative version, but are creating an open-source growing organism in more than 240 different languages.

Eruv at Sodom: honoring human diversity

Two Talmud tractates[18] relate that at the time King Solomon simultaneously established laws relating to carrying on the Sabbath (*eruv*) and washing one's hands (*n'tilat yadayim*), a heavenly voice proclaimed great joy at King Solomon's wise action. What is so significant about laws relating to carrying on the Sabbath and washing one's hands that taken together elicit the highest level of Divine rejoicing? If one were to choose two laws to express the central values of Judaism, it would seem that others would have been singled out. Most people, even non-observant Jews, know nothing about *eruv* and *n'tilat yadayim*. An *eruv* demarcates the boundary around a community within which observant Jews can carry things between their homes and the street on the Sabbath day. *N'tilat yadayim*, meaning "raising of hands," is a hand-washing ceremony performed on waking in the morning to celebrate the wonder of wakefulness. It is the first religious act of the day that is repeated throughout the day before meals and after using the toilet to sanctify one's everyday actions. The hand-washing ceremony is a private act after which our two hands are raised revealing the uniqueness of our fingerprints while reciting a blessing linking this mundane act to divine infinitude. Fingerprints symbolize individual differences; no two people have the same fingerprint patterns. Building an *eruv* is a communal act that creates community while *n'tilat yadaim* is the private act that highlights individuality.

The prohibition of carrying on the Sabbath is found in two biblical passages. The Israelites are told that they were given two portions of manna on Friday so they would not go out to the desert to gather the food and carry it home on Saturday. Some people ignored what they were told and went out to find food and found none. "Then God said to Moses, 'How long will you refuse to observe my commandments? See that God has given you the Sabbath and that is why He gives you on the sixth day a two-day portion of bread. Let every man remain in his place; let no man leave his place on the seventh day" (*Exodus 16:25–29*). In addition, it is deemed a capital offense to gather sticks on the Sabbath and carry them home (*Numbers 32–36*).

In his wisdom, King Solomon created the concept of *eruv* as an easement that makes life more pleasant on the Sabbath. An *eruv* permits a Jew to observe the law with comfort by expanding the boundaries of one's place. If the yard around a house is fenced in, it is considered to be an extension of the house. Therefore, people are permitted to carry between the house and its private yard on the Sabbath day. If two neighbors surround both their houses with a single fence then they can carry into their joint yard and into each other's houses. King Solomon proposed continuing the same process until all houses in a village are surrounded by a single extensive fence that transforms the space within it from public space to a one large private space belonging to the entire community. Minimum fencing is a series of poles connected on the top by a string in a post and lintel form known as *tzurat hapetah*, "the form of the opening." It is instructive that the open-ended thought patterns of Jewish consciousness are reflected in the structure of the *eruv*, a fence built of open forms. Today, most villages, towns, and cities in Israel have constructed an *eruv* as have hundreds of communities in the Diaspora. Entering the word "eruv" in Yahoo yields 82,700 websites (6 October 2010): Greater Boston Eruv; NW London Eruv; Sydney Eruv; Los Angeles Community Eruv; East Denver Eruv; Chicago Eruv; Eruv of Cambridge and Somerville (MA); University of Maryland Eruv; Tenafly (NJ) Eruv; Aventura (FL) Eruv; Baltimore Eruv, etc.

I showed photographs of the Miami Beach Eruv to illustrate my presentation on "The Future of Art" at the National Art Education Association convention in Miami Beach. I explained that the Miami Beach Eruv is the largest environmental sculpture in America that can be perceived as both a kinetic and a conceptual artwork. The Miami Beach Eruv is a fence that runs for twenty miles around all of Miami Beach. It carries a spiritual message while meandering through the gross material world, passing between the colorfully painted Art Deco buildings on Ocean Drive and the beautiful topless models sunning themselves on the beach. It is mainly made of poles connected by a string attached to the top of each pole in "the form of the opening." The poles are attached to palm trees and to the sides of buildings and some are free standing like telephone poles. Traffic even passes under its string lintel hovering from pole to pole over the causeways to the mainland of North America. The *eruv* changes in seven-day cycles. To observant Jews who are carrying anything, it gains the properties of a stone wall from sunset on Friday until stars appear in the sky on Saturday night. Although the *eruv* is visually transparent, it becomes conceptually opaque, impassable by a Jew carrying a book in his hand. However, it serves no halakhic purpose during the other six days of the week. It is as if it disappears. Indeed, it remains invisible all week to those who are not looking for it. None of my audience at the NAEA convention, even those living in Miami, had ever noticed the strings running from pole to pole. Although the *eruv* actually exists in space and defines it, its significance is in time. Like the Sabbath itself, the *eruv* is architecture in time.

In his poetic book, *The Sabbath: Its Meaning for Modern Man*, Rabbi Abraham Joshua Heschel emphasizes the sanctification of time in Jewish consciousness:

> Jewish ritual may be characterized as the art of significant forms in time, as *architecture in time*… The meaning of the Sabbath is to celebrate time rather than space. Six days a week we live under the tyranny of things of space; on the Sabbath we try to become attuned to *holiness in time.*[19]

After creating heaven and earth, "God blessed the seventh day, and He declared it to be holy" (*Genesis 2:3*). Of all the things that God had created, only time was divinely sanctified. In the Ten Commandments, the term "holy" (*kadosh*) is applied to one word only – the Sabbath. "God blessed the Sabbath day and made it holy" (*Exodus 20:11*). The Sabbaths are the great cathedrals of the Jewish people that it has been able to create in the lands of its dispersion. During the two millennia exile of the Jewish people from its homeland, all of its enemies have been unable to destroy its great architecture built in time. The *eruv* gains its significance marking time for those who live within it and who recognize its influence in unifying community on the seventh day of the weekly cycle.

I recognized the powerful significance of King Solomon's wisdom linking *eruv* to the hand-washing ritual of *n'tilat yadayim* while I was standing in the heat of the day at the lowest spot on Planet Earth, at Sodom, the desolate site of the notorious biblical city of ill fame that brought down God's great wrath. Ten artists were invited to participate in creating environmental artworks at the southern end of the Dead Sea, a lunar landscape where the city of Sodom once stood. It was expedient to schedule the event marking the completion of the artworks on the holiday of Purim. People from throughout Israel would be free to visit there when schools were closed and most people do not work.

Six weeks before "Sodom/Purim 5744/1984," Ezra Orion, an environmental artist from the Negev community of Sde Boker who had organized the environmental art event, arranged for the ten artists to meet at Sodom and stake out sites for their artworks. The high concentration of oxygen in the Dead Sea air amplified our sensitivities to the potent landscape as we walked about in an overpowering sauna. The artists surveyed the scene to find a site that responded to each one's intentions, thoughts and feelings. What I needed to do came to me in a flash as I stood before a hill from which two distant purple mountain ranges emerged in a haze like two wings. Reading the Bible's account gave me no clue as to what I would find actually standing there at the place where wicked Sodom once stood. The mountain range emerging in the background to the south of my Sodom hill is Edom, the biblical home of Amalek who attacked and murdered the straggling Israelites, weak from their slavery, as they trekked through the desert. Amalek is considered to be ancestor of Haman, the villain of the Purim story, who in one guise or another endures as the permanent enemy of the Jews. The range to the north is Moav, the birthplace of Ruth, progenitor of kings David and Solomon, the matriarch of the messianic line. The two mountain ranges look alike on the surface, mirror images masking differences between evil and goodness. Contemplating links between the two biblical books named for women, *Ruth* and *Esther* (the Purim story), made it overwhelmingly clear to me that my artwork should bring together Sodom and Purim by linking *eruv* and *n'tilat yadayim*. Solomon's wisdom can teach us that community symbolized by *eruv* coupled with individuality symbolized by *n'tilat yadayim* leads to the highest good when human beings create community that honors what is unique in each individual.

Sodom and Purim are linked by the bureaucratic idol of standardization that denies individuality. The *Midrash* tells us that when a traveler was unfortunate enough to seek hospitality among the Sodomites, official policy forbade, turning him away to spend the night in a forbidding wasteland. That would have been patently unforgivable. He was invited instead to enter the city and spend the night in a bed – a standard bed. If the guest chanced to be taller than average, his obliging hosts resolved the dilemma of dangling legs by cutting them off to fit the length of the bed. If he were too short, his arms and legs would be tied to a torturous mechanism that would stretch him until he fit. What was intolerable to the Dead Sea denizens was deviation from their arbitrary norm. In addition, whenever a poor, hungry man arrived in Sodom, he was never denied charity. The Sodomites gave him bricks of non-negotiable gold rather than something as pedestrian as bread. When the newly rich man, hands full gold, died of starvation, his generous hosts would retrieve their charity. It is this behavior in which the letter of the law is fulfilled while ignoring its true intentions and spiritual worth that the Talmud refers to as "acting in the manner of Sodom."

The lethal threats that erode human society converge in the stories of Sodom and Purim. Both are stories about official evil promoting intolerance of human diversity coupled with authoritarian mentality and bureaucratic rigidity. The story of Purim told in the biblical book of *Esther* records a failed attempt at government-sponsored genocide promulgated by the mighty Persian Empire nearly 2,500 years ago.

> Haman sought to destroy all the Jews that were throughout the entire kingdom of Ahasuerus [...] And Haman said to King Ahasuerus: "There is a certain people scattered abroad and dispersed among the peoples in all the provinces of your kingdom, and *their ways are different* from all the other peoples' [...] If it pleases the king, let it be recorded that they be destroyed." [...] Letters were sent by courier to all the provinces of the king, to destroy, to slay and to exterminate all the Jews in one day, from young to old, children and women, and to plunder their possessions. (*Esther 3:6, 8, 9, 13*)

The king set the wheels of his bureaucracy spinning by sealing his letter with his official signet ring, therefore, irreversibly sealing the doom of all his Empire's Jews, including unknowingly his beloved Queen Esther. When Esther rises to the occasion, reveals herself as a Jew and pleads the case of her people, it is too late for the king to reverse the damage "for an edict written in the king's name and sealed with the king's signet may not be revoked" (*Esther 8:8*). The best he could do was to issue a new edict granting the Jews of every city permission to organize and defend themselves. In his acquiesce to Haman's hatred for those who are different from the official norms and in his mindlessly issuing irrevocable edicts to destroy what is dear to him, King Ahasuerus joins the Sodomites in evil ways that remain, unfortunately, all too common in our day.

I surrounded the hill at Sodom with an *eruv* constructed from seven telephone poles connected by rope lintels and ten sawed-down poles emerging from the ridge of the hill. The short poles served as pedestals for ten one of a kind ceramic vessels filled with water for hand washing. The telephone company provided the poles and special hole-digging equipment to set them around the hill and connect them at their tops with rope in the open form, *tzurat hapetah̲*. Miriam Benjamin worked with her art students in Yeroham to make the hand-washing vessels from clay dug from Negev mountains rising to the west of Sodom. Each vessel was different in design from the others reflecting the unique vision of each student. Each of the ten short poles following the natural ridgeline of the hill to its crest was crowned with one of the ceramic vessels. From a distance, the vessel-topped poles looked like a *minyan* of people, the quorum needed to create a community of worshippers. Walking up the hill in the heat of the day, the Purim visitors looked down at the blue sky shimmering on the surface of the water in the vessels. They were pleasantly surprised when they dipped their fingers into the water and found that the water had been kept cool by its evaporation through the semi-porous unglazed pottery. I felt that my artwork on that scorching day at Sodom had created a symbolic field of energy that would reveal the inner essence concealed by the surface deception of the twin mountain ranges of Edom and Moav. Perhaps it could disarm Amalek and Haman in all their contemporary reincarnations by teaching that the highest good is reached when we create community that honors what is unique about each person. Creating community that pays tribute to the emergence of individuality and facilitates its free expression invites God's greatest joy.

Beautifying actions: adding light to the world

The traditional view of the function of art in Jewish life has been to beautify a ritual act, a process known as *hidur mitzvah*. In "Judaism and Aesthetics," Rabbi Shimon Cowen[20] proposes four categories of *hidur mitzvah*. In the first case, beauty is an inherent quality of a *mitzvah* that cannot be separated from it. It is an absolutely essential component of carrying out the precept. The Bible commands us "take for yourself *pri etz hadar*" (*Leviticus 24:40*) along with three other plant species to celebrate *Sukkot*, the Festival of Our Rejoicing. *Pri etz hadar* literally means "the fruit of a beautiful tree." *Hadar* is the adjective "beautiful" to the verb *hidur*, "to beautify." *Hadar* is related to the kabbalistic *sephirah* of *hod*, a graceful aura of splendor, to *hoda'ah*, acknowledgement of majesty, and to *hed*, reverberation of light. Before the *Sukkot* holiday in Israel, one can see people at open-air markets shopping for the four species closely examining *pri etz hadar*, the *etrog* fruit of the citron tree. The fruits, cradled in nests of flax, are laid out on tables where the buyers gently lift them close to their eyes turning the precious fruits around and around checking for flaws. Beauty determines the price of the *etrog*.

In the second category, beauty is incorporated in the body of the *mitzvah* as an addition to its basic requirements. It is a *mitzvah* to light candles on the eight-day holiday of Hanukah, the Festival of Lights. The blessing over the candles recited each night, *"l'hadlik ner shel Hanukah,"* is on kindling the Hanukah *candle*, in singular. If one does not possess enough candles, lighting one candle can fulfill the *mitzvah*. The Talmud records a difference of opinion between Hillel and Shamai. Shamai proposes lighting all eight candles on the first night removing one each night until only one remains on the last night. His argument is conceptually valid since the lighting of the candles commemorates the cruse of olive oil found to light the menorah in the Temple rededication after the Hellenistic defilers were driven out of Jerusalem. Although there seemed to be only enough oil in it for one day, it miraculously lasted for eight days. Since all the oil was in the cruse on the first day and it was used up each subsequent day until none was left after eight days, it would seem logical to follow Shamai's way. Hillel, on the other hand, proposes an opposite procedure. He proposes that we light one candle on the first night and add an additional candle each night until we light eight candles on the last night of Hanukah. Jewish tradition follows the way of Hillel where aesthetics overrides logic. It is more beautiful to add light to the world each day than remove it. Until this day, Jews have beautified the *mitzvah* of lighting the Hanukah menorah by adding more light each night.

Ritual objects, *tashmishai kedushah*, which aesthetically embellish a *mitzvah* but are extraneous to its halakhic performance, form the third category. This is the most common experience of *hidur mitzvah*. Although connected to the performance of the *mitzvah* to enhance its beauty, it is not an essential element. A beautiful silver menorah adds elegance to the celebration of Hanukah. However, kindling wicks coming out of eight Coca-Cola bottles filled with olive oil fully fulfils the *mitzvah*. A commonly used ritual object is a spice box for the *havdalah* ceremony marking the end of the Sabbath day. Blessings are made over sweet wine, a multi-wicked braided candle, and fragrant spices in a multi-sense acknowledgement that we are leaving the sacred domain of the Sabbath to enter the profane week of every day work. Augmenting the taste of the wine and the sight of the flame, we experience a pleasant fragrance to prevent the additional soul that we acquire on the Sabbath day from departing too rapidly at the close of the day. Often decorative silver spice-boxes in the forms of flowers and castle towers made for holding fragrant spices are passed around at the *havdalah* ceremony for all to smell. However, opening a bottle of cinnamon or cloves from the kitchen cupboard or picking sprigs of myrtle or geranium in the garden can serve the purpose just as well, and fulfill the *mitzvah* in full.

In the first three categories, beautification is connected to the body of the *mitzvah* in its essence or as enhancement. In the fourth category, beautification remains separate from the actual performance of the *mitzvah*. We can see the beauty of a *mitzvah* when its performance is lovingly portrayed in a painting, drawing or photograph. This category corresponds to the iconic and symbolic semiotic categories of art in contrast to the identic and dialogic semiotic categories related to the first three categories of *hidur mitzvah*. The spiritual atmosphere of a Jewish home can be enhanced by paintings depicting celebrations of lighting a Hanukah menorah or of ending the Sabbath day with wine, flames and spices.

The *mitzvah* of giving charity can be expressed artistically in each of Rabbi Cowen's four categories. I could make a painting that illustrates a scene from Theodore Herzl's utopian novel *Altneuland/Old New Land* [21] as a fourth category artwork. It would depict an elegantly dressed young lawyer putting a coin into the cold hand of a bare-footed street urchin in tattered clothes squatting in the snow at the entrance of a classy Viennese restaurant. Or I could photograph an old wrinkled woman sitting with outstretched hands on the steps leading to the Wailing Wall. Passing tourists give her coins in exchange for red threads thought to bring good luck. Both of these are examples of the fourth-level category of *hidur mitzvah*. Hanging in a home, they could be aesthetic reminders of the importance of giving charity.

The aesthetically powerful coin-collection canister that I designed to raise funds to fight leukemia expresses the third level of *hidur mitzvah*. I had seen the "Op Art" exhibition at New York's Museum of Modern Art when I was commissioned to design a coin-collection canister for the Leukemia Society of America. Inspired by the exhibition, I made a mock-up on an empty gefilte fish bottle with 1/16" black drafting tape. I taped black stripes parallel to each other running diagonally around the bottle at 1/16" intervals. This created a robust moiré pattern that shimmered with spectators' movements in true "op art" fashion. I made a more polished version on a clear plastic sheet with the black stripes running down into a band of black at the bottom with just two bold words in red Helvetica italic: *DEFEAT LEUKEMIA*. The sheets were screen-printed, wrapped around metal bases and tops with a slit for coins, and distributed throughout America. Only filling the canister with coins could put an end to the disturbing, disorienting sense of movement created by the shimmering moiré pattern. My coin collection canisters were very successful in enticing people to do the *mitzvah* of giving charity. The "Defeat Leukemia" canister was a modern version of the historic silver coin-collection canisters with intricate baroque decorative relief designs seen in most Jewish museums. This represents the third category of embellishing the *mitzvah* of giving charity.

The first two categories in which the artwork is an integral part of the *mitzvah* are described at the end of this chapter. I describe *cybersight@tikunolam*, an artwork that uses digital technologies to translate visual into tactile perception so that blind people can "see" pictures on a computer screen by moving a specially designed computer mouse that traces the pictures on their fingertips. Corporate sponsors contribute funds for medical research to fight blindness each time an image of things blind people said they would most like to see are contributed by web surfers to the artwork Internet site.

LightsOROT at MIT: learning Torah through art

Complimenting Rabbi Cowen's four categories, I propose that a fifth category is the *hidur* of the particular *mitzvah* of Torah study, a *mitzvah* considered to be equal to all the other *mizvot* combined. Creating art can be an alternative method of Torah study that beatifies the *mitzvah* of study through visual *midrash*. *Midrash* is the unique Jewish literary form that combines commentary, legend, and narrative explanations of biblical texts. In a sense, *midrash* fills the spaces between the written words to reveal deeper meanings of scriptural passages. Art as visual *midrash* provides fresh commentaries on biblical texts through multimedia experiences that extend the verbal exploration of text into visual realms. The centrality of learning Torah in Jewish life warrants its being an additional category.

Arthur Danto, in his book on the visual arts in post-historical perspective[22], expounds on the major significance of *context* in understanding postmodern art. Rabbi Arthur Green reads "context" in its primal meaning "with-text."

> Judaism is a process of ongoing commentary. To be a Jew is to be a student. To be a self-affirming Jew is to love and study Torah. It is no small matter that the rabbis considered study equal in value to all the other *mitzvot* combined as one […] It is, after all, that historical *contextuality*, that living "with the text" that places us within the chain of tradition, that makes our generation a contributor to the sum of what will be passed on to the future. We do this faithfully only as we submit ourselves to the role of student, *as we are willing to allow ourselves to be shaped by the text as we have received it and made it our own.*[23]

The centrality of the *mitzvah* of learning Torah is revealed in the daily morning liturgy in response to the special blessing for Torah study:

> "Please *YHVH*, our God, sweeten the words of Your Torah in our mouth and in the mouth of Your people, the family of Israel. May we and our offsprings and the offsprings of Your people, all the House of Israel know your Name and study your Torah for its own sake."

Following the blessing, three passages from the Written and Oral Torah are studied. The first is the biblical blessing, "May God bless you and safeguard you. May God illuminate His countenance for you and be gracious to you. May God turn His countenance to you and establish peace for you" (*Numbers 6:24–26*). The last is a passage from the Talmud:

> These are the precepts whose fruits a person enjoys: honor due to father and mother, acts of kindness, early attendance at the house of study morning and evening, hospitality to guests, visiting the sick, providing for a bride, escorting the dead, absorption in prayer, bringing peace between man and his fellow – and the study of Torah is equivalent to them all.[24]

Creating art as a method of Torah study is exemplified by the creation of *LightsOROT* at MIT's Center for Advanced Visual Studies. *LightsOROT* (*OROT* is the Hebrew word for "lights") is a collaborative artwork that resulted in a communal creation that honors and facilitates the expression of each individual artist's vision and talents. Creating community that honors what is unique in each person is a core Jewish value that we learned from King Solomon's instituting *eruv* with *n'tilat yadayim.* The very process of creating *LightsOROT* is as much an expression of Jewish consciousness as it is creating visual *midrash* on light. *LightsOROT* exemplifies a reconstructive postmodernism that moves beyond modernism's construction of personality cults built upon the individual artist's originality and postmodernism's deconstruction of modernism's values. Creating *LightsOROT* formed a community of artists who worked together to create an integral environmental artwork that revealed each artist's vision in interplay with the visions of the others.

LightsOROT focuses on studying the subject of "light" in Torah and natural systems through artistic explorations that exploit new light media. The Bible soundly links light to Torah study: "A *mitzvah* is a lamp and Torah is light" (*Proverbs 6:23*). The biblical passage read in response to the blessing for Torah study talks of "illumination." Seeking creative avenues for coaxing divine light to enter our everyday world and illuminate it forms the core of Jewish consciousness. This inspired search that integrates postdigital, halakhic, kabbalistic, ecological and dialogic perspectives through artists' collaborative study and creative expression forms the essence of postmodern art. *LightsOROT* is an exemplary model of the process of creating art through Torah study that resulted from my forging a creative partnership between MIT's Center for Advanced Visual Studies and Yeshiva University Museum, between America's most advanced center for exploring new technologies in creating art and America's leading center for Jewish learning.

The genesis of *LightsOROT* can be traced to my Torah study with Moshe Davidowitz (now Dror). I contacted Moshe when I read that he was creating a Center for Art in Jewish Life in New York in 1974. At that time, I was art professor at Columbia University and Moshe was directing the continuing education program at New York University. Moshe had formerly been a congregational rabbi in Connecticut and director of the Spertus Museum, the Jewish museum in Chicago. Our shared interests prompted us to become a *h̲evrutah*, a dyad of Torah learning partners, meeting alternatively at Columbia and NYU. We began our studies at the beginning with the book of *Genesis* integrating our study of traditional Jewish sources with insights gleaned from the natural and social sciences, general systems theory, art history, aesthetic philosophy, and media ecology. Light became a central focus of our learning Torah together. My Torah studies with Moshe paralleled a graduate course on color that I was teaching at Columbia where my students explored color from the points of view of physics, physiology, psychology, anthropology, aesthetics, and art history. I had written the book *Light and Sight*[25] during my earlier career as a science teacher.

I moved back to Israel in 1977 where I founded a new regional college in the Negev desert town of Yeroham in cooperation with Bar-Ilan University where I was a professor. I spent the summer of 1980 in the States as a research fellow at MIT's Center for Advanced Visual Studies. During that summer, I began my dialogue with the Center's director, Otto Piene. Otto was cofounder with Heinz Mack of Group Zero[26] in Germany and pioneered in creating light art. The first public performance of his *Light Ballet* was in 1959. We had talks about the future of light art as an expression of the spiritual dimensions of the electronic revolution. My researches that summer led to the first stages in the development of a biofeedback-generated computer graphics system that provided the digital technology for my artwork, *Inside/Outside: P'nim/Panim*, that was to become part of *LightsOROT* eight years later.

That summer, I met Moshe Dror in New York. He had moved from a houseboat in the Hudson River while at NYU to a villa overlooking Lake Geneva to head the International Humanistic Psychology Association. I invited him to join me in Yeroham as dean of the college when he finished his term of office in Geneva. Before Rosh Hashanah the following year, he phoned me to tell me that he would be accepting my invitation. Moshe and I, rabbi and artist, resumed and intensified our *ḫevrutah* study of Torah in relationship to space-time concepts, sense modalities, new digital technologies, and alternative roles for the arts in Jewish life. We put our ideas into practice by developing a program in art and Jewish thought at the college in which we both taught and coauthored a paper, "Educating a Jewish Artist,"[27] on the theoretical background for our new program. It was during our long walks together in the desert mountains over the next few years that the conceptual framework for *LightsOROT* evolved. It became clear to us that God/Torah/Israel were intimately linked to light in Jewish consciousness: "God is my light" (*Psalms 27:1*); "Torah is light" (*Psalms 6:23*); Israel is a "light unto the nations" (*Isaiah 42:6*). Light is connected by *gematria* to the secret of the Infinite. *Or* (light) and *raz* (secret) and *ein sof* (infinite) all share the numerical equivalent of 207.

When Sylvia Herskowitz, Director of the Yeshiva University Museum, was visiting Israel, Moshe and I drove up to Jerusalem to propose to her developing an art exhibition exploring Jewish thought through electronic technologies. She was enthusiastic about the idea.

After seven years living in the desert and teaching at the university, I earned a sabbatical year that I spent at MIT. My renewed dialogue with Otto Piene made me realize that the conceptual framework forged in Israel could be transformed in physical reality with access to the powerful technological resources of MIT. Otto became my new learning partner in the further conceptualization of *LightsOROT* in relation to artistic collaboration and electronic technologies. I flew down to New York to meet with Sylvia Herskowitz and she subsequently made visits to Cambridge to meet with Otto and me in planning *LightsOROT* as a joint venture between MIT's Center for Advanced Visual Studies (CAVS) and Yeshiva University

Museum. As the idea of the show became a reality, Otto and I expanded our dialogue to include other artists affiliated with CAVS. We organized a symposium on light in Jewish thought and experience at Yeshiva University in New York for the artists who would be collaborating in creating *LightsOROT*. Moshe Dror flew in from Israel to participate in the symposium. All the artists, both Jews and non-Jews, spent two intense days learning with Moshe and me.

After completing my year as research fellow at CAVS, I accepted the position to head the art department at Pratt Institute in New York. By moving to New York and becoming a frequent flyer on the New York–Boston shuttle, I was able to coordinate a complex collaborative enterprise between the YU museum and CAVS. During the next two-and-a-half years, I met regularly with Otto and the participating artists both in group meetings and in individual conferences. I worked closely with all the artists in exploring interfaces between the conceptual framework for the show and their individual talents and aesthetic concerns. The paper that I wrote for the collaborating artists, "Lights of Creation: Sight and Insight," became their conceptual guidebook. I invited rabbis with wide-ranging creative insights to CAVS to engage in dialogue with the artists about interrelationships between Torah concepts and ideas for their artworks. Being on campus, Daniel Shevitz, the director of Hillel, the Jewish students organization at MIT and organizer of a conference on computers and Judaism, was available for consulting with the artists. Zalman Schachter, the spiritual leader of the Jewish Renewal movement and professor at Temple University who learned Torah in Dharamsala with the Dalai Lama[28] learned with the artists. Josef Wosk, a congregational rabbi in Massachusetts and graduate of both the rabbinical school of Yeshiva University and Harvard Divinity School, not only learned with the artists, but he collaborated with Joel Ziff in preparing "A Light Concordance," a resource document on light in Jewish sources that became a 14-page section of the *LightsOROT* catalog.

My dialogue with Otto expanded to involve three other artists who worked with us in solidifying fluid ideas into a concrete set of visual and auditory parameters that developed into a master plan for integrating all the collaborating artists' work within the museum space. My dialogue with musical composer and laser artist Paul Earls and environmental artist Lowry Burgess had begun during my 1980 summer at MIT, during subsequent visits to Cambridge, and at the "Sky Art" exhibition in Munich. Lowry had also visited with me in Israel where I assisted him in distilling the waters from the world's major rivers at the Dead Sea for the core of his art satellite. Joe Moss, sculptor, environmental artist, and art professor at the University of Delaware, joined us and was named project director responsible for the construction and installation of the exhibition at the museum.

Talit Firmament was the first of 25 pieces to be installed. It was created by Beth Galston, a sculptor working with light to create participatory environments. She collaborated with textile designer Maura Walsh and laser artist Paul Earls. Beth had created sensitive environments that involve movement and change with scrim cloth used in theater sets. Since she had been working with cloth, seeing the white woolen *talit* prayer shawl with black stripes became an inspiration that led her to spread layers of black and white scrims over the entire ceiling of the museum. In dialogue with Paul Earls, it became a firmament on which he could project laser animations of shimmering Hebrew words alternating with celestial luminaries drawn by Otto Piene. He was inspired by the biblical account in *Genesis 1:6–8, 14*:

> God said, "Let there be a firmament" ... God made a firmament ... God called the firmament "Heaven" ... God said, "Let there be luminaries in the firmament of the heaven to separate between day and night. And they shall serve as signs".

He created a computer program that moved a tiny mirror at extremely high speeds to send a coherent beam of laser light onto the scrim firmament synchronized with an environmental sound composition. Paul had earned three degrees from Eastman School of Music at University of Rochester, including a Ph.D. in composition. He used an argon laser beam to trace contour-line drawings of dancing light forms projected through layers of black and white scrim. Looking up, the visitor to *LightsOROT* saw stars emerging from the firmament and Hebrew letters spelling out the words, *b'reshit* (in the beginning), *or* (light), *rakeeyah* (firmament), *yom* (day), and *lailah* (night), in response to words spoken by visitors. Microphones hanging from the ceiling captured spoken words transmitting them into a computer, processing them, and then returning them to the exhibition space as shimmering words hovering above, projected through layers of scrim cloth.

Immediately after reading the Ten Commandments in the Bible, we read, "All the people *saw* the sounds" (*Exodus 20:15*) rather than "heard the sounds." Paul was able to simulate the synaesthetic experience at the foot of Mount Sinai of seeing sounds. When the word "*or*" was said, ten-foot letters *alef, vav, resh* spelled out the Hebrew word for "light" in pure laser light flickering across a scrim night sky for all to see. In his artist's statement for the *LightsOROT* catalog, Paul Earls wrote that he strives "to create new work that neither I nor anyone else has ever experienced before which integrates the senses and celebrates life."[29] He followed his statement with a passage from *Psalms 19:2*: "The heavens declare the glory of God and the firmament showeth His handiwork."

Beth's image of *talit* called for *tzitzit* fringes. On the four corners of the scrim, I attached giant *tzitzit* tied from ship rope given to the museum by the Israeli shipping firm, Zim Lines. Instead of using rope for one of the strands of each fringe, I used light-transmitting fiber optic cable used in telecommunications. I painted it sky blue in accordance with the biblical precept, "put on the fringe of each corner a blue thread" (*Numbers 15:38*). Each blue fiber optic bundle was frayed at the ends into 6,000 dots of light as memorial lights for the six million Jews murdered by the Nazis. To produce six millions dots of light, I would have had to make four thousand giant *tzitzit*! Harold Tovish's sculpture *Region of Ice* was also a memorial light for those lost in the Holocaust. Light emerged through a heap of broken glass like an icy mass of shattered lives. A lone life-size sculpted head entombed in an epoxy sphere rested on the broken glass.

When I looked up at the scrim *talit* and firmament, I began to see them from another perspective, as a big tent. The scrims were transformed into the patriarch Abraham's tent pitched in the desert inviting passersby to study Torah there with him. "He planted an *eshel* in Beersheba and there he proclaimed the Name *YHVH*, God of the Universe" (*Genesis 21:33*). Although *eshel* is usually translated as a tamarisk tree, *midrash* reads alternative meanings into it. *Eshel*, spelled *alef*, *shin*, *lamed*, becomes the word for "question" *sha'al*, when the order of the same three Hebrew letters is rearranged as *shin*, *alef*, *lamed*. The Talmud[30] teaches that *eshel* is also the acronym for food, lodging, and escorting guests on their way. Abraham built an inn, a learning retreat in the desert where people could enter and ask him questions about his new monotheistic idea. The inn was a tent open to the four *ruhot*, both winds and spirits. *Ruhot* refers both to winds blowing into the tent from the north, south, west, and east and to the spiritual individuality of each human being coming to Abraham's open school from different places in their intellectual and emotional lives. Like the four sons in the Passover *Haggadah*, each represents a different personality type whose different questions require different answers. After learning with Abraham, each individual could go in his own way leaving through one of the four openings appropriate to him. In addition to providing food (*alef*) and lodging (*shin*) in the dormitory of his school, Abraham would escort (*lamed*) each of his students on his way into the desert until the student felt secure enough to continue asking questions on his own. *Lamed* is both the name of a letter and a word in itself related to learning, study and teaching. It is the tallest of the 22 Hebrew letters, is the only one that breaks through the scribe's guide line scored above the letters. It soars above all the letters into a mysterious space where no other letters have been. Learning is posing questions evoked by this mystery. *Eshel* and *sha'al* (question) are one, written with the same Hebrew letters.

The never-ending process of asking questions according to Albert Einstein is the essence of both art and science as they approach the mysteries of human life and nature. "He who knows it not and can no longer wonder, no longer feel amazement, is as good as dead, a snuffed-out candle."[31] When Nobel laureate in physics, Isador Rabi, was asked how he was able to reach such heights in science, he explained that as a child all the mothers of his classmates would ask, "What did you learn in school today?" while his mother asked, "Did you ask any good questions today?" Jewish consciousness relates scientific curiosity to spirituality. When Moses first found God in the desert, he was drawn by curiosity about an anomalous physical phenomenon. A bush was burning and was not being consumed. It was not in a mystical trance or in a holy place that Moses found God, but in researching a lowly shrub. Moses is instructed to take off his shoes and move aside to see the bush from a fresh vantage point in order to discover the deep significance of his encounter.

There is a dispute recorded in the Talmud about the symbolic meaning of the *eshel* that Abraham planted in Beersheba. Although some of the rabbis saw *eshel* as an inn in which Abraham taught, others explained that *eshel* is a generic name for trees. It symbolizes a grove, a *PaRDeS*, a vertical metaphor for learning Torah in depth from multiple vantage points in contrast to the horizontal metaphor of a tent/inn/school open in four directions to engage individuals with different cognitive styles and worldviews in inquiry-centered learning. *PaRDeS*, as we discussed earlier, is the acronym for four levels of learning Torah.

P'shat is the simple, literal meaning of the text. ***Remez*** is a hint of deeper significance. ***Drash*** is homiletic interpretation. And *Sod* is the mystical, hidden meaning.

Although Beth's work on the tent-like structure symbolized Abraham's open school, I felt that we needed a complimentary piece that would more explicitly convey the idea and feeling of entering and exiting from the tent/inn/school. While discussing this need with Otto, he was fingering a sand-cast glass hand that I had seen over the years on his desk. I digressed to add the idea that the Hebrew word for "knowledge" *yadah* is made up of the word for "hand" *yad* and the letter "*ayin*" which is also the word for "eye." To know is to see what your hand is doing, to be aware of one's behavior, and to take responsibility for one's actions. I returned to talking about how Abraham comes to see by changing his position both physically and psychologically, walking away from all that is familiar. "Walk in your own way away from your land, your birthplace, from your father's home to the land that you will recognize when you see it" (*Genesis 12:1*). As I began exploring connections between Abraham's walking, "*lekh l'khah*," and Torah learners walking in and out of Abraham's tent/inn/school, Otto exclaimed: "Wouldn't it be beautiful to sand-cast glass feet like this glass hand on my desk. Paths to enlightenment could spiral out in four directions." We had found the conceptual and aesthetic counterpoint on the floor below to the scrim firmament above. Feet in sand brought to mind the vision of Abraham's descendents walking through desert sand from Egypt to the foot of Mount Sinai where they received the Torah. It is said that the centuries of slavery made Abraham's descendents forget the Torah that Abraham knew intuitively and, therefore, they had to be handed it in writing at Mount Sinai.

Walking through the halls of Massachusetts College of Art where my wife was studying for her MFA, I had seen the flames from kilns running non-stop to maintain batches of molten glass for the students. With the help of the MassArt faculty and students joined by our MIT students, we cast 88 glass feet in sand. Men, women, and children pressed their bare feet into damp sand. Molten glass was then poured into the impressions left by their feet. We built a raised plywood floor a foot above the museum floor and painted it black. Four paths of 22 glass feet were placed on the plywood floor spiraling out from the center of the museum's main exhibition hall. Pencil lines were drawn around the feet on the plywood and foot-shaped openings in the false floor were cut out with a saber saw. After the glass feet were glued with epoxy into the corresponding openings, a separate light bulb was wired under every one of the feet. Each of the four paths contained 22 illuminated feet corresponding to the 22 letters of the Hebrew alphabet with which the Torah is written. All of the lighting in *LightsOROT* emanated from the ambient light of the artworks themselves. No other lighting was added.

The four sets of illuminated feet emerged from under a large satellite dish (13 feet in diameter) facing upwards. A second identical satellite dish facing downwards hovered several feet above it. Mirrored plastic sheets cut into 70 narrow triangles were glued into each dish like pie slices. Rotating in midair between the two dishes was a prism star cast in acrylic. Under the bottom dish an infrared sensor sensed the body heat of a visitor as she approached the dishes. The sensor triggered a flash of intense light on her face that was captured by the prism star and spread onto the 70 mirrored triangles. She saw her face replicated 70 times as a symbolic expression of the Talmudic saying that there are "*ayin panim l'torah*" (70 faces of the Torah). Since *ayin* is both "70" and "eye," the Talmudic saying encourages us to see every passage in the Torah from 70 different viewpoints. The electronic flash is a metaphor for the sudden flash of insight (*ẖokhmah*) that brings us a fresh viewpoint. A second version of this dialogic artwork stood vertically inviting the visitor to stand between the two dishes. He would see his face 70 times circulating around animated laser drawings of eyes floating in the circular opening in the center of each satellite dish. Both versions of this interactive environmental sculpture were a truly collaborative work of Lowry Burgess, Paul Earls, Christopher Janney, Joe Moss and Otto Piene.

To fine-tune its interactive proper ties, the vertical version, *Monocle 2*, was exhibited at the Museum of Science in Boston before it was installed at the Yeshiva University Museum. Media artist Vin Grabill made a video of Otto and me talking about *LightsOROT* standing between the dishes at the science museum. This became the conceptual introduction for a video catalog of the show that Vin produced. At the Yeshiva University Museum, a continuous loop of the video catalog was shown in a room with rows of seats off the main exhibition hall. We also produced a 113-page print catalog that explores the theoretical background of light in Jewish consciousness from multiple viewpoints. It was dedicated to Gyorgy Kepes, Head of the Light and Color Department at the New Bauhaus in Chicago and Founding Director of the Center for Advanced Visual Studies at MIT, who was a great inspiration for many of the artists participating in *LightsOROT*. In addition to essays by Otto and me, world-renowned psychologist of art Rudolf Arnheim wrote the introductory essay, "On the Symbolism of Light." Professor Arnheim was teaching at Harvard at the time and had written *Toward a Psychology of Art*, *Visual Thinking*, and *Entropy and Art*. For the catalog, I also coauthored with Rabbi Norman Lamm, Torah scholar and President of Yeshiva University, "Light, Vision and Art in Judaism: A Dialogue." Rabbi Lamm's book, *Torah Umadda:The Encounter of Religious Learning and Worldly Knowledge in the Jewish Tradition*,[32] has provided intellectual support for my life-long quest for understanding interrelationships between art, science, and Jewish thought that was artistically realized through the creation of *LightsOROT*.

Suspended from the ceiling beneath the scrim firmament, a 60-foot-long water prism ran across the entire length of the museum to create *River of Light*. In Hebrew, river and light are linguistically linked. The feminine form of word for "river" *nahar* is a biblical word for "light" *naharah* used in *Job 3:3*. Lowry Burgess, Joe Moss, and David Burns bent 30-inch acrylic sheets into a "V" shape that was suspended with steel cables from a concrete beam in the museum ceiling. Filled with water, the 60-foot prism weighed several tons. A bank of light projectors projected white light through the full length of the water prism to create a two-foot-high spectral band that colored the length of the museum wall opposite the light source. We found that this massive water prism acted as a seismograph responding to tremors across the globe that caused the undulation of the spectral band. Seeing this seismographic effect provided an awesome aesthetic experience of ecological perspective in the realization that our entire planet is interconnected as a single system that touches me. Rainbow waves generated by earthquakes and volcanic eruptions throughout the world as revealed by an artwork in a New York museum evoked a spiritual response in many people. The concept that white light is composed of a spectrum of different colors has both kabbalistic and halakhic significance. Just as white light breaks into colors, the one divine light breaks into the spectrum of *sephirot* in its descent into our every day world. *Sephirot* as stages in the creative process can be represented by a continuum of spectral colors. It is also a metaphor in Jewish tradition for community that is beautiful when it recognizes the wonderful variety exhibited by the individuals who make up the community. The white light of community and the range of spectral colors representing individuals are one.

The spectrum assigned to Hebrew letters can reveal hidden patterns in the Torah. With the computer programming assistance of Yisroel Cohen, I created a dialogic artwork, *Torah Spectrograph*, through which people could see these patterns as related to their own lives. Just as each Hebrew letter has a numerical value, it can also have a color value. The first letter of the alphabet, *alef*, with the numerical equivalent of the all-inclusive one, is represented by one pixel of white light. The second letter, *beit*, is two pixels of red light. The third letter, *gimel*, is a band of three letters of orange. When we come to the eighth letter, *h̲et*, a band of eight violet pixels, we have run out of spectral colors. We then repeat the spectrum three times so that the final letter of the alphabet, *tav*, is a band 22 violet pixels long. To access the Torah color patterns, the visitor enters his birthday into the computer from which the date of his birthday in the Hebrew calendar is calculated. That Hebrew date determines what portion of the Torah is read in synagogue each week. At age thirteen, a Jewish boy is called to the Torah to acknowledge his becoming a bar mitzvah. At age twelve, a Jewish girl becomes a bat mitzvah. The Hebrew date of the visitor's birthday calls up his or her Torah portion and plays it out on the monitor in bands of color according to the spacing pattern in a Torah scroll. The *Torah Spectrogram* addresses each individual with a personalized biblical symphony of rainbow colors scrolling across the computer monitor. I have subsequently used the same digital schema in more high touch materials. In the courtyard of a Miami synagogue, I wrote out *Psalm 146* "from generation to generation" in bands of stained glass in a wedding canopy. With the sun passing through the canopy, we see a biblical song of color moving over the white wedding gown of the bride as the Earth spins. I also spelled out the story of the creation of the universe from *Genesis* in bands of wood painted in acrylic colors flowing across the desert surface in the Negev mountains.

A spectral image appeared as visitors moved before the computer-generated white light hologram, *The Ten Commandments: Stone of Light*. I made photocopies of the two versions of the Ten Commandments as they appear hand-written in a Torah scroll, one version made by God and the second made by Moses. In collaboration with two scientists working in Steve Benton's MIT laser lab, artist/holographer Dieter Jung used my photocopies to create a three-dimensional hologram of the two versions superimposed on one another. Visitors experienced the scribe's Hebrew letters floating in space as pure disembodied spectral light. Rabbi Chaim ben Atar in his book *Or Hachaim* (Light of Life), wrote that the Ten Commandments were written with a divine finger of light.

Wen-Ying Tsai, an artist with degrees in engineering, created *Dancing Menorah.* He constructed his five-foot-high menorah with delicate stainless steel rods painted in spectral colors emerging from the seven branches of the menorah. The movement of the rods was controlled by an audio control system coordinated with stroboscopic lights to realize the biblical passage: "All the people *saw* the sounds" (*Exodus 20:15*) rather than "heard the sounds." Sounds made by the visitors caused the long slim rods to sway in dancing motion that was visually stopped periodically by stroboscopic light.

A grid of hologram discs created by Yaacov Agam was hanging on the balcony surrounding the main exhibition hall. Visitors looking through them at any light source below saw the point of light transformed into a glowing six-pointed star. When looking down on the 88 illuminated glass feet through Agam's holograms, the visitor would see a glowing Star of David hovering over each of the footprints.

Todd Siler created *Metaphors on Light*, multimedia bookworks using enamel, Plexiglas, oilstick, canvas, acetates, filmstrips, inks, paper, and adhesives integrated through collage and assemblage. His three bookworks draw titles for the kabbalistic books, *Sepher Habahir* (Scroll of Brightness), *Zohar* (Glow), and *Orot* (Lights). Todd's artwork explores inner realms of brain activities in relation to light in Jewish experience. He explains that as books of light they are timeless reflections on the celestial working of the human mind and the boundless elements of the imagination. I first met Todd in 1980 when I was research fellow at MIT and he was a graduate student there. He had lived in Israel and had illustrated the Passover *Haggdah*, *Gates of Freedom*, in delicate watercolor paintings. Our common interests in a biological model of creative process in relation to art, technology, and culture made our dialogue particularly animated. After completing his master's degree at CAVS, he studied neurophysiology and cognitive psychology and became the first artist to be awarded a Ph.D. by MIT. He expanded his doctoral research in his book, *Breaking the Mind Barrier: The Artscience of Neurocosmology.*[33] His work truly expresses the ecological perspective of Jewish consciousness.

The biofeedback computer graphics system that I created in collaboration with my son, Ari, for our *Inside/Outside: P'nim/Panim* dialogic artwork also deals with links between neurophysiology and psychology. As described earlier in the book, this artwork creates a feedback loop in which one's internal mind/body state (*p'nim*) changes a video image of one's external self (*panim*). It is an interactive system in which visitors paint digital self-portraits with the flow of their inner rivers of light. An individual sits before a monitor and video camera with one finger in a plethysmograph, a device that measures blood flow in the finger and is a sensitive indicator of mind/body state. We developed special software that transformed the color, form, and size of one's face on the monitor in response to changes in mind/body states. A portrait derived from Jewish consciousness is a dynamic changing system presenting the flow of life forces between spiritual and material realms rather than a static painting in which one's still face, enclosed in a gold frame, is set off from the world of life. It is living life that is honored, not still life, not *nature morte*.

Poet Elizabeth Goldring and video artist Vin Grabill collaborated on an interactive video art work, *Eye/Sight*, which brought visitors inside their own eyes and the eye of a person whose vision is being lost. A video camera reaches the retinal screen of a visitor's eye and juxtaposes it with a medical examination of Elizabeth's eye. Elizabeth explored the visual disintegration of her own eyes as it produces visible inference patterns from electromagnetic fields of light and color. She says,

> I want to grasp these patterns as they rattle my soul, as they are shocked by maelstroms of sunlight, laser light, and blood light. For me in the terror and beauty of these moments words/images/sounds/space are exploded and fused.[34]

An audiotape of Elizabeth's *Eye/Sight Poems* could be heard.

During a discussion with Elizabeth and Otto, Elizabeth mentioned her collection of rooster crows from around the world. It immediately bought to mind the morning prayers that begin with the blessing, "Blessed art Thou, Lord our God, King of the Universe, who hast given the rooster intelligence to distinguish between day and night." This blessing is derived from *Job 38:36* where the Hebrew word for rooster (*sekhvi*) comes from the verb "to see." The rooster sees the first light of dawn and announces it to us. Otto jumped at the idea of creating a large inflatable rooster that sings the song of dawn for *LightsOROT*. With the assistance of Lees Ruoff, Otto created a giant *Sekhvi Rooster*. Entering the museum, the visitor saw a massive heap of red nylon cloth laying on the floor in darkness. It suddenly began to move, grow, and come to life as it was inflated to its full 17-foot height. As it was growing, the rooster was illuminated by strobe lights that made the growth occur at an erratic pace. Strobe lights stop motion. The lighting was designed by MIT's Harold "Doc" Edgerton, the inventor of the strobe light. His famous strobe photographs of a milkdrop crown, a bullet blasting through an apple, flying birds with their wings' motion stopped, have been extensively reproduced. When fully inflated, the lighting danced in tune to *sekhvi* roosters' crows circling the globe. The inflated rooster first called out a long stretched-out crow recorded in Japan where they have contests won by the length of the rooster's crow. Crows from roosters in the Land of the Rising Sun are followed by the "cock-a-doodle-do" of American roosters, rooster crows from across Europe, roosters announcing the dawn from the hills of Jerusalem with "coo-coo-ree-coo", crows from India and China, and finally the long crows of Japanese roosters again. The blowers then stopped, thereby, causing the *Sekhvi Rooster* to slowly deflate and collapse into a dark heap before new breath would enter it and resurrect it. Augmenting Elizabeth's collection of rooster crows, crows of champion Japanese roosters were collected by art critic Itsuo Sakane and Jerusalem rooster crows were recorded by my son, Ron.

Ron also collaborated with Paul Earls on the sound environment for *LightsOROT*. Ron is a rabbi and a scientist who teaches biology and Torah and relationships between them at the Yeshiva High School for Environmental Studies at the edge of the Ramon Crater in the Negev. At the time, he was a student at Merkaz Harav, the rabbinical college founded by Rabbi Kook in Jerusalem, as well as a music student. He recorded ten different ways of chanting from the Torah scroll of ten different Jewish communities in Israel. Ten people chanted in the tradition of their fathers the first four days of creation (*Genesis 1:1–19*), from "Let there be light" on the first day to "Let there be luminaries (*orot*) in the firmament" on the fourth day. The diverse tunes of Yemenite, Kurdish, Persian, Dutch, German, Polish, Hungarian, Eastern European and American Jews became the raw materials for Paul's musical composition. Using computers, voice synthesizers, and other digital technologies, he wove together a four-channel musical environment linked to his laser display on the scrim firmament. He spaced four amplifiers and speakers in the museum to create a sound environment in which there was a continual interplay between the different Torah chanting styles. His musical composition made so many combinations and variations from the ten styles that it took a full year before the same combinations reappeared. For the 18 months that *LightsOROT* filled the museum, visitors were encircled and engaged by a rich sound environment in which alternative musical traditions in the biblical Hebrew language were transformed through digital technologies and linked to a laser display of celestial luminaries on the scrim firmament. They listened to distinctive traditional melodies in their cultural individuality as they joined in harmony with each other to create new voices for our postdigital era in keeping with Rabbi Abraham Isaac Kook's proclamation, "Renew the old, and sanctify the new."

Responsive art in a postdigital world

"We need an art that transcends the distanced formality of aesthetics and dares to respond to the cries of the world."[35] These words are the major thesis of Suzi Gablik's book on the future of art, *The Reenchantment of Art*. *Cybersight@tikunolam* is an artwork that attempts to respond to the cries of the world by transforming kabbalistic perspective into halakhic perspective through digital technologies. It extends the concept of *hidur mitzvah*, beautifying a deed, from deeds between human beings and God (*bein adam v'makom*) to deeds between people (*bein adam v'havero*). It a responsive artwork that extends the *sephirah* of Beauty (*tiferet*) into the *sephirah* of compassion and loving kindness (*hesed*) through activities that engender life enhancement and creative renewal. It reaches out to human beings lacking the primary sense required to encounter art as defined by Western culture. *Cybersight@tikunolam* offers blind people opportunities to experience imagery through their sense of touch using unique digital technologies developed in Jerusalem. They can gain tactile access to those things they would most like to see and to images from their everyday life. Through the Internet, access is extended globally to the blind as websurfers contribute images that generate funds for research to fight blindness.

Cybersight@tikunolam is the embodiment of Gablik's vision that "the next historical and evolutionary stage of consciousness, in which the capacity to be compassionate will be central not only to our ideas of success, but also to the recovery of both a meaningful society and a meaningful art."[36] This next stage is the postdigital transition from Hellenistic to Hebraic consciousness. "When art is rooted in the responsive heart, rather than the disembodied eye, it may even come to be seen, not as a solitary process it has been since the Renaissance, but as something we do with others."[37]

Cybersight@tikunolam is responsive art that gives eyes to the blind and hands to art. Art of the past may have expressed social and humanitarian concerns, but it hangs insularly on a museum wall disengaged from the issues that define it. In a sense, that art is handicapped. It possesses no hands to help the cause it is advocating. Responsive systems art plugs art into the real world transforming its audience into active participants. It has hands to reach out and invite people to collaborate in fixing the world, *tikun olam*. When art has hands for receiving and giving, art gains a soul.

The genesis of *cybersight@tikunolam* was a discussion with my son, Ari, about extending into the social realm the individual experience at the human–machine interface of our bioimaging artwork, "Inside/Outside: *P'nim/Panim*," that we had created at MIT. Our work at MIT led us to explore how art interfaces between real space and cyberspace by creating a cybernetic feedback loop between them. We began brainstorming about how actions in cyberspace could effect changes in people's lives in real space, how the Internet can bring people together to help one another, how digital technologies can be used for *tikun olam*, fixing the world by filling it with loving kindness, and how web art could actually generate charity. We were not interested in making a painting depicting one person giving a coin to another or designing a beautiful silver coin collection canister to aesthetically enhance the experience of giving charity, we sought ways to move beyond making art *about* compassion and charity, beyond illustration and decoration, to creating art in which actually performing acts of compassion and charity provide the aesthetic experience. Ari suggested that he build a website in which people worldwide would be invited to contribute pictures to the site. Like the funding of walkers in a walkathon, we could get corporate sponsors to donate money to a charity each time an image is contributed. We ended our discussion at his home in New Hampshire not knowing what kinds of images we would request, what charities would benefit, what corporations would be appropriate sponsors, and in what venues would the completed artwork be presented to the public.

On my returning home to Israel, my wife suggested that a solution to our questions could be found by adding a second *mitzvah* to the *mitzvah* of giving charity. Miriam turned the negative *mitzvah* from the Torah, "You shall not put a stumbling block before the blind," (*Leviticus 19:14*) into a positive one: "Restore sight to the blind so they will not stumble." Her suggestion came from the heart since she has retinitis pigmentosa, a genetic defect that has made her legally blind. The artistic team had expanded to three – Miriam, Ari and me. As the artwork evolved, our collaboration expanded to include more and more people. *The Third Hand: Collaboration in Art from Conceptualism to Postmodernism*[38] argues that collaboration in art is the crucial element in the transition from modernism to postmodernism. The redefinition of art and artistic collaboration intersected during this transition.

We began by asking people who were born blind or became blind at a young age: "What are four things that you would most like to see if you had vision?" We interviewed blind people in Israel, the Czech Republic, and the United States and sent questionnaires worldwide to associations and schools for the blind. We have received responses from countries as disparate as Australia, Ethiopia, Fiji, India, Korea, Lebanon, Lithuania, Niger, Poland, Slovenia, Zambia and the United Kingdom. The similarity of responses from such diverse cultures teaches us about the common vision of humanity. Blind people worldwide wanted mostly to see those things they could not touch: the sky, the sun, moon, stars, clouds, and airplanes and landscapes, mountains, lakes, forests, the seashore and ocean waves.

Ari created a website on which we posted the results of our cross-cultural research. Web surfers were invited to contribute pictures of things that blind people most want to see. In response to these contributed images, corporate sponsors contribute funds for scientific research to prevent and cure blindness. We created a *cyberpushka*,[39] a postdigital age coin-collection box of global dimensions, an artwork through which actions in cyberspace rectify defects in real space, *tikun olam*.

The next stage is to link the Internet to innovative digital technologies that enable blind people to "see" pictures through the sense of touch. A special computer mouse was developed in Jerusalem to give blind people direct access to pictures on a computer monitor. Beneath fingers placed in indentations in this specially designed mouse, there is a grid of pin-like protrusions that move up and down independently to trace the image on the computer monitor onto the blind person's fingertips. I drove up to Jerusalem to meet with Dr. Roman Guzman who is the inventor of this digital system. Dr. Guzman, a cognitive psychologist, and I discussed ways that his technology could develop aesthetic experiences for blind people. Using this new technology, blind people worldwide could access pictures from the image bank at our website.

Complementing the *cyberpushka* part of the artwork, we gave blind people cameras to take pictures of their environments and experiences. Instead of framing a picture through the viewfinder, they point the cameras in response to sounds, smells, touch, movement and intuition. In pilot trials in Tel Aviv, Prague and Miami, we found that the images created by blind people present fresh and original viewpoints. Sighted photographers, from rank amateurs to seasoned professionals, frame their images through a viewfinder. Unframed images shot by blind people are less stereotypic. They provide alternative perceptual approaches that are often found to be aesthetically powerful by sighted viewers.

Each blind photographer discussed her images with us. The most conceptually and visually interesting photographs were scanned and digitized. Using a computer graphics program, I transformed the photographs into line drawings that were rendered as tactile hard copy. The computer images were transferred to a swell paper imbedded with a chemical that makes a raised line appear when heated. The blind photographers could perceive their pictures by running their fingers over the protruding lines. They also could access their photographs directly from the computer using Guzman's tactile mouse. It is an act of *hesed*, compassion and loving kindness, to facilitate a blind person's "seeing" images of her own world through her fingertips. It corresponds with the prophetic vision: "the eyes of the blind will be opened," "from darkness and blackness the eyes of blind people will see." (*Isaiah 35:5, 29:18*)

The website exhibition in cyberspace can become museum exhibitions in real space. In *010101: Art in Technological Times*, David Ross writes:

> We recognize that there are at least two very distinct kinds of exhibition space available to artists today and that, while we recognize and respect their differences, clearly they can be deployed in parallel in service of a single project.[40]

Cybersight@tikunolam takes different forms when exhibited on the Internet and in spaces bounded by museum walls. In museum settings, sighted visitors can see photographs taken by blind people mounted next to the computer-generated line drawings. Both blind and sighted visitors can run their fingers over relief hard-copy versions of these same images as if to "see" them through their sense of touch. This tactile experience is unavailable in virtual space. However, in both real or virtual spaces a blind person can sit in front of a computer monitor and actively participate by using the tactile mouse to trace the computer images on her fingertips.

Art actively responding to the cries of the world is most exalted in Jewish tradition. Rabbi Nachman of Breslov teaches:

> The key to the workings of the entire universe is charity. The entire flow of spiritual and physical blessing into the world can be seen as God's "charity" to His creatures, to bring them to know Him. When we ourselves give charity, we are participating in this process, which is why charity is such an exalted mitzvah.[41]

Judaism teaches that the four letters of the divine name *YHVH*, *yud, hei, vav, hei,* contain the mystery of charity. *Yud* is the smallest letter of the Hebrew alphabet that symbolizes a coin. Since *hei* is the fifth letter of the alphabet with the numerical value of five, it alludes to the five fingers of the hand. The first *hei* is the hand that gives. The letter *vav* is shaped like an outstretched arm. When used as a prefix, *vav* is the conjunction "and" – the connecting word. *Vav* connects the one who gives to the one who receives. The final *hei*, is the hand of the one who receives. Indeed, the very name of God visually teaches about the centrality of charity in Jewish life. The Internet greatly expands the connective power of the *vav*, an outstretched arm reaching out to people across the global network and linking them to one another while generating charitable contributions to fund research efforts to fight blindness.

The Lubavicher Rebbe, Rabbi Menachem M. Schneerson, teaches that the sweeping technological changes we are experiencing today were predicted some two thousand years ago in the *Zohar*, a classical text of Jewish mysticism. It describes how the outburst in scientific knowledge and technological advancement would be paralleled by an increase in sublime wisdom or spirituality. Integrating the wisdom of the mind and the wisdom of the soul, which is the role of the artist, can begin to usher true unity into the world.

Notes

1. Soloveitchik, *Halakhic Man*, p. 99.
2. Schneerson, *Torah Studies*, p. 320.
3. *Babylonian Talmud: Bava Metzia*, 59b.
4. Soloveitchik, *Halakhic Man*, p. 81.
5. Ibid., p. 30.
6. Ibid., p. 15.
7. David Avraham Spector, *Shu't Omanut* [Responsa on Art] (Jerusalem: Emunah College and Erez, 2003). In Hebrew.
8. Nehama Leibowitz, *Studies in Vayikra*, trans. Aryeh Newman (Jerusalem: World Zionist Organization, 5740/1980), p. 178.
9. *Babylonian Talmud: Yoma*, 85b.
10. Benjamin, "The Work of Art in the Age of Mechanical Reproduction."
11. Soloveitchik, *Halakhic Man*, p. 31.
12. *Babylonian Talmud: Shabbat*, 30a.
13. *Pirke de Rabbi Eliezer* 36.
14. Gideon Weitzman, *Sparks of Light* (Northvale, NJ, and Jerusalem: Jason Aronson, 1999), pp. 16–17.
15. *Pirke de Rabbi Eliezer*, quoted in Meir Zlotowitz (translation and commentary), Bereishis/Genesis (Brooklyn, NY: Mesorah, 1986), p. 337.
16. Samson Raphael Hirsch, *The Pentateuch* (New York: Judaica Press, 1986), pp. 54–58.
17. Halprin, *Taking Part: A Workshop Approach to Collective Creativity*.
18. *Babylonian Talmud: Eruvin*, 21b, and *Shabbat*, 14b.
19. Abraham Joshua Heschel, *The Sabbath: Its Meaning for Modern Man* (New York: The Noonday Press, Farrar, Straus and Giroux, 1993), p. 8, 10, original emphasis.
20. Shimon Cowen, "Judaism and Aesthetics," *Journal of Judaism and Civilization*, vol. 3, pp. 85–108.
21. Theodore Herzl, *Altneuland/Old New Land*, trans. Lotta Levensohn (Princeton: M. Wiener, 1997). Original German edition published in 1902.
22. Danto, *Beyond the Brillo Box: The Visual Arts in Post-historical Perspective*.
23. Arthur Green, *Seek My Face, Speak My Name* (Northvale, NJ and London: Jason Aronson, 1992), p. 138, original emphasis.
24. *Babylonian Talmud: Shabbat*, 127a.
25. Melvin L. Alexenberg, *Light and Sight* (Englewood Cliffs, NJ: Prentice-Hall, 1970).
26. See Otto Piene and Heinz Mack, *Zero* (Cambridge, MA: The MIT Press, 1973). Original German edition published in 1958.
27. Menahem Alexenberg and Moshe Dror, "Educating a Jewish Artist," *Bisdeh H̲emed*, No. 5–6, May–June 1984. In Hebrew.
28. For the story of Rabbi Schachter's encounter with the Dalai Lama, see: Rodger Kamenetz, *The Jew in the Lotus* (San Francisco: Harper, 1994).

29. Alexenberg and Piene, *LightsOROT*, p. 85.
30. *Babylonian Talmud: Sota* 10a.
31. Einstein, *The World As I See It*, p. 5.
32. Norman Lamm, *Torah Umadda: The Encounter of Religious Learning and Worldly Knowledge in the Jewish Tradition* (Northvale, NJ, and London: Jason Aronson, 1990).
33. Todd Siler, *Breaking the Mind Barrier: The Artscience of Neurocosmology* (New York: Touchstone Books, Simon and Schuster, 1990).
34. Alexenberg and Piene, *LightsOROT*, p. 91.
35. Gablik, *The Reenchantment of Art*.
36. Ibid., p. 182.
37. Ibid.
38. Green, *The Third Hand: Collaboration in Art from Conceptualism to Postmodernism*.
39. *Pushka* is theYiddish word for the traditional coin canister for collecting charity.
40. David A. Ross, *010101: Art in Technological Times* (San Francisco: San Francisco Museum of Modern Art, 2001), p. 012.
41. Rabbi Noson, leading pupil of Rabbi Nachman of Breslov, *The Fiftieth Gate*: *Likutey Tefilot*.

Index

About the Author

Mel Alexenberg is an artist, educator, writer, and blogger working at the interface between art, science, technology, and culture. His artworks explore interrelationships between postdigital narratives and Jewish consciousness, space-time systems and electronic technologies, participatory art and community values, high tech and high touch experiences, blogart and wikiart, and responsive art in cyberspace and real space. His artworks are in the collections of more than forty museums worldwide from the Metropolitan Museum of Art in New York to the Jewish Museum of Prague to the Israel Museum in Jerusalem.

Alexenberg is head of the School of the Arts, Emuna College, Jerusalem, and professor emeritus at Ariel University. He was professor of art and education at Columbia University and Bar Ilan University, head of the art department at Pratt Institute, dean of visual arts at New World School of the Arts in Miami, and research fellow at MIT's Center for Advanced Visual Studies.

His books include: *Educating Artists for the Future: Learning at the Intersections of Art, Science, Technology, and Culture* (Intellect Books/University of Chicago Press), *Dialogic Art in a Digital World: Judaism and Contemporary Art* (in Hebrew), *Aesthetic Experience in Creative Process* (Bar Ilan University Press), and with Otto Piene, *LightsOROT: Spiritual Dimensions of the Electronic Age* (MIT and Yeshiva University Museum). He has contributed chapters to the books: *Digital Visual Culture, Interdisciplinary Art Education, Semiotics of Visual Culture: Sights, Signs, and Significance* and *Community Connections: Intergenerational Links in Art Education.*

He lives with his wife, artist Miriam Benjamin, in Petah Tikva, Israel.

Website: www.melalexenberg.com
Blog: http://artiststory.blogspot.com

www.ingramcontent.com/pod-product-compliance
Ingram Content Group UK Ltd.
Pitfield, Milton Keynes, MK11 3LW, UK
UKHW051912100926
8556IPUK00006B/857

9 781841 5037